Field Identification of Minerals
For Oregon Prospectors

DOGAMI Bulletin 16

Compiled by the staff of the
Oregon Department of Geology and Mineral Industries

This work contains material that was originally published in 1940 by
the State of Oregon with the assistance of the United States Geological Survey.

Introduction

It has been nearly seventy five years since the Oregon Department of Geology and Mineral Industries released it's important publication "Field Identification of Minerals For Oregon Prospectors". First released in 1940, this important volume has now been out of print for years and has been unavailable to the mining community since those days, with the exception of expensive original collector's copies and poorly produced digital editions.

It has often been said that "*gold is where you find it*", but even beginning prospectors understand that their chances for finding something of value in the earth or in the streams of the Golden West are dramatically increased by going back to those places where gold and other minerals were once mined by our forerunners. Despite this, much of the contemporary information on local mining history that is currently available is mostly a result of mere local folklore and persistent rumors of major strikes, the details and facts of which, have long been distorted. Long gone are the old timers and with them, the days of first hand knowledge of the mines of the area and how they operated. Also long gone are most of their notes, their assay reports, their mine maps and personal scrapbooks, along with most of the surveys and reports that were performed for them by private and government geologists. Even published books such as this one are often retired to the local landfill or backyard burn pile by the descendents of those old timers and disappear at an alarming rate. Despite the fact that we live in the so-called "Information Age" where information is supposedly only the push of a button on a keyboard away, true insight into mining properties remains illusive and hard to come by, even to those of us who seek out this sort of information as if our lives depend upon it. Without this type of information readily available to the average independent miner, there is little hope that our metal mining industry will ever recover.

This important volume and others like it, are being presented in their entirety again, in the hope that the average prospector will no longer stumble through the overgrown hills and the tailing strewn creeks without being well informed enough to have a chance to succeed at his ventures.

Kerby Jackson
Josephine County, Oregon
July 2014

INTRODUCTION

PURPOSE

Need for a bulletin of this type is shown by the numerous inquiries that are received requesting information on the identification of rocks and minerals. These inquiries indicate that such information should include:

1. Mineral identification information written especially for Oregon prospectors and collectors.
2. Minerals described should be common in Oregon, or if not common, should be of economic importance.
3. Tests for minerals that involve the use of materials at hand; blowpipe, chemical, and microscopical tests should be omitted because the necessary equipment is seldom with the man in the field.
4. A low publication price to permit wide distribution.

This bulletin is compiled from various texts on mineralogy and from the experiences of the Department's staff. The terminology is as non-technical as possible, and where technical terms are used, they are explained in the text or glossary. Only simple mechanical aids to identification (field equipment) are suggested, to avoid the necessity of chemical and blowpipe tests. Those who wish more detailed descriptions and technical discussions may find such information in various books on mineralogy (see selected bibliography). The United States Geological Survey and the United States Bureau of Mines have many publications dealing with specific minerals and containing pertinent suggestions concerning the economics of mining and marketing.

This bulletin is one of a series designed to provide authentic information about the mineral resources of the State of Oregon, to indicate what to seek, to give instruction on the identification of minerals and how to develop and finance a mineral deposit, as well as to outline the economics of marketing the product.

THE PROSPECTOR AND THE COLLECTOR

Several groups may find this bulletin or another of this series of value. The prospector desires to identify minerals that have a possible commercial value. The collector also wants this information and, in addition, data on how to make a collection of minerals for his own enjoyment and financial benefit. The recreationist wishes general information that will permit him to enjoy his surroundings just as he wishes to know the names of flowers and trees, but has no desire to start a herbarium, or a mineral collection.

ASSAY AND MINERAL IDENTIFICATION SERVICE

One purpose of the Oregon State Department of Geology and Mineral Industries is to help residents become more familiar with the rocks and minerals occurring in the state by identifying samples submitted to the Department. A short summary of the services relating to the analysis of ores and minerals is given here.

Assays

The Department will make two free assays of ore or mineral samples for the common metals in any thirty day period for any citizen or group of citizens of the state provided the samples are collected within the state. Each sample must be accompanied by a properly filled-in information blank (fig.2).

The material or sample to be assayed should be representative, that is, a channel sample. A channel sample consists of the material taken from a channel of uniform width and depth across a vein or formation containing the ore or mineral.

Samples for assay should be sent to the nearest State Assay Laboratory. These are located at Grants Pass and Baker.

Mineral Identification

Free mineral identification service is provided citizens of the state at the Baker and Grants Pass laboratories and also at the Portland office. Specimens sent in will be classified, and the enclosed minerals will be identified and described, if possible, and the commercial importance and uses of the material represented will be indicated.

An information blank (fig.2) or letter should accompany each specimen, giving location (the place of collection), an estimate of the amount of material present, and such other specific information about the specimen as the sender can supply.

Qualitative Tests

The Department will be pleased to make qualitative tests for the common metals like lead, zinc, copper, chrome, manganese, etc., and encourages prospectors and collectors to have these tests made. Tests for gold and platinum require regular fire assays. Because of the time and expense involved, the Department does not ordinarily make fire assays unless the samples are properly taken or cut from a ledge.

The Department does not have the facilities to test for the rare metals, such as beryllium and gallium. Few assay laboratories do have these facilities. Such samples should be sent to some laboratory which has adequate spectroscopic equipment.

Sample information blanks are provided by the Department for your convenience. The "Law Relating to Free Analysis of Ores, Minerals, etc.", is listed on the back of figure 2.

At all three of the Department's offices (Portland, Baker and Grants Pass) mineral collections are maintained for study purposes. These collections are open to the public during office hours.

MINERALS DEFINED

A mineral is defined as a "natural inorganic substance and, <u>when pure</u>, has a definite chemical composition, usually a definite crystal form and specific physical properties, such as cleavage, fracture, color, hardness, luster, and specific gravity". This bulletin is concerned chiefly with the physical properties.

Examination of minerals shows that they usually crystallize in definite geometrical forms when conditions permitted. For example, quartz forms crystals in cavities.

Unfortunately, from the standpoint of simplicity, minerals are not always pure, and their chemical composition and crystal form may vary within certain narrow limits. Their variations usually are easily detected by use of tests for physical properties and will not be discussed herein.

Although tests for chemical composition are not described in this bulletin, the chemical composition is given for the reader's information. This composition is usually written as a series of symbols - a sort of chemical shorthand. For example, the formula of calcite, $CaCO_3$, indicates that one atom of calcium, one atom of carbon, and three atoms of oxygen are used to form the $CaCO_3$ molecule. The list of "Radicals" given in the table includes a few groupings of elements and it will be noted that CO_3 is carbonate. So the mineral is calcium carbonate. The following table may serve as an aid in interpreting this shorthand. (See Figure 1, following page).

ACKNOWLEDGEMENTS

Data contained in this bulletin have been taken from the standard texts on mineralogy, and supplemented with facts and ideas supplied by the staff of the State Department of Geology and Mineral Industries. Especial acknowledgement is made to Earl K. Nixon, Leslie M. Motz, F. W. Libbey, and A. M. Swartley, for their careful criticism of the manuscript and many helpful suggestions, and particularly to H. B. Wood and Wayne R. Lowell for carefully checking the tables and mineral descriptions.

Dr. Lloyd W. Staples, of the University of Oregon, has carefully corrected and amended this revised second edition.

Figure 1.

ELEMENTS AND THEIR SYMBOLS

Actinium	Ac	Indium	In	Samarium	Sm
Aluminum	Al	Iodine	I	Scandium	Sc
Antimony (stibnium)	Sb	Iridium	Ir	Selenium	Se
Argon	A	Iron (ferrum)	Fe	Silicon	Si
Arsenic	As	Krypton	Kr	Silver (argentum)	Ag
Barium	Ba	Lanthanum	La	Sodium (natrium)	Na
Beryllium (glucinum)	Be	Lead (plumbum)	Pb	Strontium	Sr
Bismuth	Bi	Lithium	Li	Sulfur	S
Boron	B	Lutecium	Lu	Tantalum	Ta
Bromine	Br	Magnesium	Mg	Tellurium	Te
Cadmium	Cd	Manganese	Mn	Terbium	Tb
Calcium	Ca	Masurium	Ma	Thallium	Tl
Caesium	Cs	Mercury (hydrargyrum)	Hg	Thorium	Th
Carbon	C	Molybdenum	Mo	Thulium	Tm
Cerium	Ce	Neodymium	Nd	Tin (stannum)	Sn
Chlorine	Cl	Neon	Ne	Titanium	Ti
Chromium	Cr	Nickel	Ni	Tungsten (wolframium)	W
Cobalt	Co	Nitrogen	N	Uranium	U
Columbium (niobium)	Cb	Osmium	Os	Vanadium	V
Copper	Cu	Oxygen	O	Xenon	Xe
Dysprosium	Dy	Palladium	Pd	Ytterbium	Yb
Erbium	Er	Phosphorus	P	Yttrium	Y
Europium	Eu	Platinum	Pt	Zinc	Zn
Fluorine	F	Polonium	Po	Zirconium	Zr
Gadolinium	Gd	Potassium (kalium)	K		
Gallium	Ga	Praseodymium	Pr		
Germanium	Ge	Protoactinium	Pa		
Gold (aurum)	Au	Radium	Ra	Common Radicals	
Hafnium (celtium)	Hf	Radon (niton)	Rn	Ammonium	(NH_4)
Helium	He	Rhenium	Re	Carbonate	(CO_3)
Holmium	Ho	Rhodium	Rh	Phosphate	(PO_4)
Hydrogen	H	Rubidium	Rb	Hydroxyl	(OH)
Illinium	Il	Ruthenium	Ru	Sulphate	(SO_4)

ROCKS DEFINED

A rock is an aggregate of minerals, so it is necessary to be able to identify minerals before much can be done with rock determination. One definition of a rock states that it is "any naturally-formed mass of mineral matter, whether or not coherent, constituting an essential and appreciable part of the earth's crust." Thus, quartz is a mineral, but should it form a large and appreciable mass it may be a rock. For example, sandstone is an aggregate of quartz grains cemented together by some kind of natural cement such as silica, lime carbonate, iron oxide, etc A complete description of various rock types is given in the bulletin on prospecting.

STATE DEPARTMENT OF GEOLOGY AND MINERAL INDUSTRIES

	1069 State Office Building	
2033 First Street	Portland 1, Oregon	239 S.E. "H" Street
Baker, Oregon		Grants Pass, Oregon

REQUEST FOR SAMPLE INFORMATION

The State law governing analysis of samples by the State assay laboratory is given on the back of this blank. Please supply the information requested herein fully and submit this blank filled out along with the sample.

Your name in full_____

Street or P.O. Box_____City & State_____

Are you a citizen of Oregon?_____Date on which sample is sent_____

Name (or names) of owners of the property_____

Are you hiring labor?_____Are you milling or shipping ore?_____

Name of claim sample obtained from_____

Location of property or source of sample (If legal description is not known, give location with reference to known geographical point.)

County_____ Mining District_____

Township_____Range_____Section_____Quarter section_____

How far from passable road?_____ Name of road_____

	Channel (length)	Grab	Assay for	Description
Sample no. 1	_____	____	_____	_____
Sample no. 2	_____	____	_____	_____

(Samples for assay should be at least 1 pound in weight)

(Signed)_____

___DO NOT WRITE BELOW THIS LINE - FOR OFFICE USE ONLY - USE OTHER SIDE IF DESIRED___

Sample Description_____

_____

Sample number	GOLD		SILVER				
	oz./T.	Value	oz./T.	Value			

Report issued_____Card filed_____Report mailed_____Called for_____

SIR-5

LAW RELATING TO FREE ANALYSIS OF ORES, MINERALS, ETC.
CHAPTER 179, Section 10, Oregon Laws 1937.

The department shall make, or cause to be made, quantitative determinations of ores and minerals when submitted for the purpose, that are from original prospects or properties within the state of Oregon, and shall mail to the sender the results obtained within 10 working days after receipt of samples; said service shall be performed by the department without charge to the sender and shall be rendered in exchange for information for the records of the department, stating the name and residence of the sender together with a history of the ore or mineral, giving as nearly as possible the location from which the sample was taken, including the name of the county, and any other matters that may be beneficial touching the same. All determinations shall be performed under the following rules and regulations and subject to the following restrictions:

(a) No sample submitted by engineers sampling prospects or mines for the purpose of evaluation, or submitted by operating mines milling or shipping ore, or hiring labor, shall be accepted by the department for assay and/or analysis, except they be taken in the field by members of the department's staff in conducting the work of the department within the scope of this act.

(b) The number of samples which any single person or group of persons may submit shall be limited to two in any 30-day period and all samples shall be assayed or analyzed by the department in the order received, as far as possible.

(c) All information received and results of determination sent out shall be open to public inspection and may be published by the department.

(d) Before any work is done on the material submitted, all information required must be possessed by the department and the 10-day limit for reports will count from the time such data are received by the department.

EQUIPMENT

Advice on equipment is given as an answer to the many requests for such information. Cost is always important and it should be remembered that with careful buying one usually gets just about what he pays for. A good workman should have good tools. Cheap tools may suffice for a time but usually prove to be more expensive. Individual preference and need vary, and for this reason many beginners use less expensive equipment until they can decide from experience what they prefer. Used equipment frequently can be acquired, but the purchaser should know his materials and be certain of getting what is represented.

FIRST AID KITS

A good first aid kit to be used in the field should be included with every outfit. Any physician or reliable druggist can give excellent advice on the purchase of such a kit. The contents should be checked at frequent intervals to make sure that the materials are in good condition. A first aid manual also is important; the "Manual of First-Aid Instruction" issued by the U. S. Bureau of Mines, obtainable from the Superintendent of Documents, Washington, D.C., for 20¢, is recommended.

A snake-bite kit, either of the hypodermic type which is expensive, or the "suction pump" type which is less expensive, is important and very effective. A person bitten by a poisonous snake should remain as quiet as possible. Stimulants should be avoided as they tend to accelerate the distribution of the poison thru the system. For this reason a physician's advice should be received before alcoholics are administered.

Spotted fever from tick bites is a danger anywhere east of the Cascade Mountains. Antitoxic serum shots may be obtained from a physician or from a Public Health Service and should be taken a month before going into the field. It is effective only for one year. Early spring is the time when tick poison is particularly dangerous. In event the serum is not available, careful examination of the entire body should be made at least twice daily and any ticks removed with tweezers. Ticks take 2 to 4 hours to settle down to serious biting.

Before going into the field one should have his general physical condition checked by a physician to determine whether he is susceptible to any particular diseases or ailments. Such action may save others from being inconvenienced, or actually endangered, by one's inability to withstand the rigors of trips.

Training by a competent instructor in methods of administering first aid is valuable.

HAND LENS

A hand lens, or pocket magnifier, is one of the collector's and prospector's best friends. The magnifying power should be about 10, and the lens surface should be protected with some sort of case.

The type of hand lens should be adapted to the individual user. The most popular type is the "doublet" or "triplet" lens, as it has a very short focus, requiring the lens close to the object and the eye close to the lens.

Single Lens Magnifying Glass

The single lens magnifying glass (fig.7) is the inexpensive "reading glass" type, but its magnification is seldom over four diameters (dimensions increased four times). It is not recommended for mineral identification.

Magnifiers in Vulcanite Cases

These magnifiers (fig.5) consist of single magnifying lenses mounted in hard black rubber cases, and magnify about four diameters. Other models have two and even three of these lenses fastened together in such a manner that they can be used singly or together, giving magnifications of four to twenty diameters (fig.5). The cost varies from $1 to $2.

Coddington Magnifier

The Coddington lens (fig.3) is a cylinder of glass, the ends of which are spherically-ground so that the unit behaves as a lens. The cylinder is enclosed in a metal or vulcanite case. The lens is quite efficient, but the "field" (area visible) is small. A lens of 10 diameters magnification should cost about $3.50-$4.00.

Doublet Magnifier

The general appearance of the doublet magnifier is similar to that of the Coddington lens (fig.4). However, the optical glass consists of two simple lenses mounted in a case. The field is much flatter than the field of the Coddington lens, that is, the visible area in focus is larger, and the magnification may range from 6 to 20 diameters. Doublet magnifiers sell for about $3.00-$3.50.

Triplet Magnifier

The triplet magnifier is much like the doublet except that it has three lenses instead of two, and more care is used in its manufacture. Much distortion is eliminated and the field is quite flat. Magnification is similar to that of the doublet type. The cost ranges from $4.00 to $7.50.

Coat Pocket Microscope Magnifier

The coat pocket microscope magnifier (fig.8) is really a small microscope that may be held in the hand. There is a small mirror at the lower end to reflect light into the tube and give a brighter image. Magnifications of 40, 50 and 60 diameters are common, but at these higher magnifications any unsteadiness of the hand or body produces vibrations that make vision difficult. The cost is about $5.00.

HAMMER

If the hand lens is the prospector's and collector's best friend, a hammer is his "rod and staff". A common type is the prospector's pick (fig.6) having a hammer and a sharp pointed pick on opposite ends of the head. Certain models have a steel shank that is continuous with the head and extends through the entire length of the handle.

Prospector's picks may be obtained at any first-class hardware store at a cost varying from $1.50 to $3.50.

KNIFE

A husky knife, fairly large to withstand severe use, is an important accessory. The blades should be made of good steel and kept well sharpened. One blade should be magnetized to serve as an emergency magnet. A leather-reamer or punch blade will be found useful for testing the relative hardness of minerals. Knives of the "Boy Scout" type are very handy; the extra "gadgets" will more than repay the gibes leveled at such an instrument.

MAGNET

A magnet is necessary to test the magnetic attraction of certain minerals, and to remove black sand (magnetite) from gold pannings. The horseshoe type is most satisfactory. A new magnet, made by the Crucible Steel Company, has more magnetism and therefore more power than the ordinary kind. It costs about 50¢.

GOLD PAN

A gold pan, with its flat bottom and widely flaring sides, is valuable for concentrating the heavier minerals (fig.9). Some pans are made of polished sheet steel, some of aluminum, and others are of sheet steel with copper bottoms for amalgamating gold. The standard-sized plain iron gold pan 16 inches in diameter is recommended. It sells for about $1.10. A smaller size, 12 inches in diameter, is convenient, especially when weight is an item. Prices vary from 60¢ to $1.00 for the iron pans.

STREAK PLATE

A streak plate is a piece of unglazed porcelain, usually about 2" x 3" x $\frac{1}{4}$" thick, used to detect the color of mineral powders. It costs about 15¢ and unless broken will last a number of years. The rough side of interior-trim tiles, or the broken edges of pottery and stoneware, serve the purpose of a streak plate.

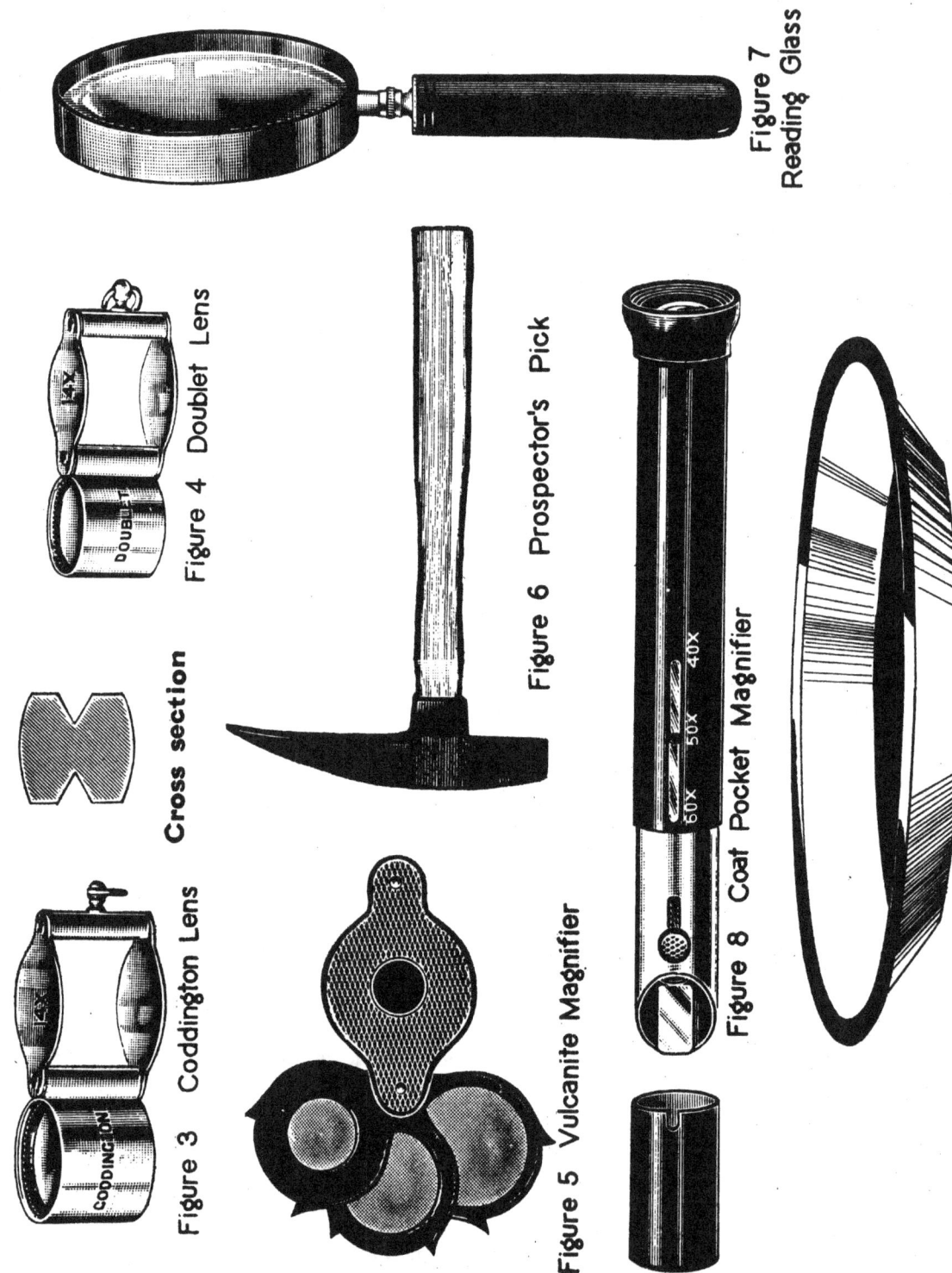

Figure 7
Reading Glass

Figure 4 Doublet Lens

Cross section

Figure 3 Coddington Lens

Figure 6 Prospector's Pick

Figure 5 Vulcanite Magnifier

Figure 8 Coat Pocket Magnifier

Figure 9 Gold Pan

ACID BOTTLE

An acid bottle is a glass bottle having a ground-glass stopper, and preferably having a capacity of about two ounces. Hydrochloric muriatic acid diluted in the proportion of two parts of acid to three parts of water is customarily used although sulfuric acid, diluted 1 to 4, may be used. In an emergency, strong vinegar may serve the purpose of a weak acid. Nitric acid, diluted with one part acid to one part water, is used for cleaning gold dust and separating gold from amalgam.

MAPS

Good maps are essential to the prospector. Those commonly available are described below.

Forest Maps

Maps of the National Forests are issued in two scales, 1/4 inch to the mile, and 1/2 inch to the mile. They show roads, trails, streams, cultural features, and are known as planometric maps. They may be obtained, free of charge, from the Forest Supervisor of the particular National Forest, or from the office of the Regional Forester, U. S. Forest Service, Portland, Oregon. Figure 10 shows areas covered by these maps.

Topographic Maps

The United States Geological Survey has undertaken to map the entire United States and show topography by means of contour lines. About half of Oregon is covered in this manner. These maps are printed as quadrangles. The scales are 1/2 inch to the mile, covering an area 24 miles by 36 miles, and 1 inch to the mile, covering an area 12 miles by 18 miles. These maps are quite accurate. They may be obtained from stationery stores or by writing the U. S. Geological Survey in Washington, D.C., and forwarding 10¢ for each map desired. Figure 11 shows the quadrangles in Oregon which were available in July 1939.

County Maps

County maps and township ownership maps, costing about $1.00 each, have been prepared by Metzger Map Company, Swender Blue-print Company, Hansen Blueprint Company, and others, of Portland. County engineers of many counties have prepared county road maps, copies of which they sometimes sell. The cost will average about $1.00 each.

Oil Company Road Maps

Road maps issued free by the various large oil companies are useful, particularly in showing the main roads and the mileages between towns, but are too small-scaled for detailed use.

12.

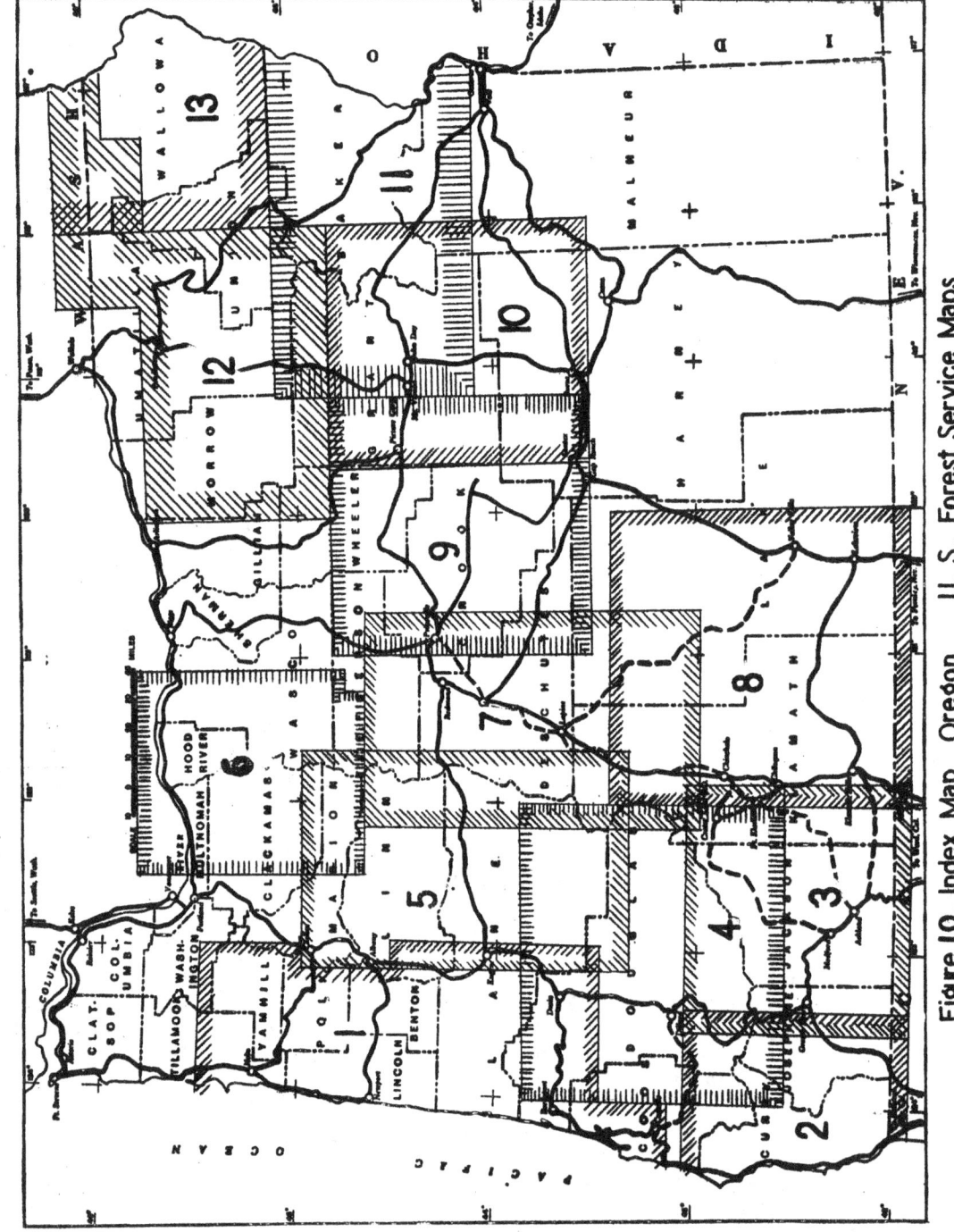

Figure 10. Index Map Oregon U. S. Forest Service Maps

TOPOGRAPHIC MAPS OF OREGON 1958

15-Minute Quadrangles
(Scale 1:62,000)

Name	Location	Date
Abbott Butte	N-8	1947
Agness	O-2	1956
Albany	G-6	1944 A
Aldrich Mountain	H-21	1943
Alsea	H-4	C1942 A
Anlauf	K-6	1939
Ashland	Q-8	1954
Astoria	A-3	C1939
Bandon	M-1	1944
Bates	G-24	1951
Battle Ax	F-10	1956
Birkenfeld	B-5	1955
Blachly	I-4	C1942 A
Blaine	D-4	1955
Blue River	I-9	1955
Bonanza 1	P-14	B
Bonanza 2	P-13	B
Bonanza 3	Q-13	B
Bonanza 4	Q-14	B
Bone Mountain	N-3	1954
Boring	D-9	1944
Bridal Veil	C-10	1954
Brownsville	H-7	1952
Burns 1	J-22	B
Burns 2	J-21	B
Butte Falls	O-8	1954
Cactus Mountain	B-32	B
Calimus Butte	O-12	A
Camas	C-9	1941 A
Camas Valley	M-4	1955
Cannon Beach	B-3	1955
Canyon City 1	H-24	B
Canyon City 3	I-23	B
Canyon City 4	I-24	B
Canyonville	N-5	1956
Cape Blanco	N-0	1956
Cape Ferrelo	Q-1	1956
Cape Foulweather	F-2	1944 A
Cascadia	H-9	A
Cathlamet	A-5	1941
Cave Junction	Q-4	1954
Cherryville	D-10	1955
Chetco Peak	Q-3	1954
Chiloquin	O-11	A
Chucksney Mountain	J-10	1955
Clatskanie	A-6	1952
Collier Butte	P-2	1954
Colton	E-9	1955
Coos Bay	L-2	1945
Copperfield	F-31	B
Coquille	M-2	1945
Cornucopia	E-30	A
Corvallis	G-5	C1942 A
Cottage Grove	J-6	1921 A
Courtrock	G-21	1956
Crow	J-5	1945
Culp Creek	K-7	1955
Cuprum	E-32	B
Dale	F-23	1951
Dallas	F-5	C1942 A
Days Creek	N-6	1954
Dead Horse Butte	B-31	B
Desolation Butte	F-24	1951
Detroit	G-10	1956
Diamond Lake	M-10	1956
Dixonville	M-6	1954
Drain	K-5	1954
Dufur 1	D-14	B
Dufur 2	D-13	B
Dufur 3	E-13	B
Dufur 4	E-14	B
Durkee	G-29	A
Dutchman Butte	N-4	1948
Eagle Cap	E-29	1954
Eagle Rock	I-16	1948
Echo Mountain	H-10	A
Elk Mountain	C-30	A
Elkton	K-4	1955
Elmira	I-5	C1942 A
Empire	L-1	1944
Enright	C-4	1955
Enterprise	D-29	A
Euchre Mountain	F-3	1957
Eugene	I-6	1949
Fairdale	D-5	1955
Fairview Peak	K-8	1955
Fish Creek Mountain	E-10	1956
Flora 1	B-30	B
Forest Grove	C-6	A
Fort Klamath 2	N-11	A
Galice	O-4	1948
Garwood Butte	M-9	1954
Glendale	O-5	1954
Glide	L-6	1954
Gold Beach	P-1	1954
Gold Hill	P-6	A
Goodwin Peak	J-3	1956
Grand Ronde	E-4	1955
Grants Pass	P-5	1954
Grass Valley	D-15	B
Halfway	F-30	A
Halsey	H-6	1941 A
Hardesty Mountain	J-8	1955
Harl Butte	D-31	1954
Harney 2	J-23	B
Hebo	E-3	1955
Heceta Head	I-2	1956
He Devil	D-32	1922 A
High Rock	E-11	1956
Hillsboro	C-7	1943 B
Homestead	E-31	A
Hood River	C-12	A
Hood River 3	C-11	B
Huntington	H-29	1951
Hyatt Reservoir	Q-9	1955
Illahee Rock	L-8	1955
Imnaha	C-31	1954
Ivers Peak	L-3	1955
Izee 3	I-21	B
Izee 4	I-22	B
Jamieson	I-29	1950
John Day	H-23	1943
Joseph	D-30	A
Kalama	A-7	C1943
Kernan Point	C-32	1954
Kimberley	F-20	1953
Klamath Falls 2	P-11	B
Klamath Falls 3	Q-11	B
Klamath Falls 4	Q-12	B
Klamath Marsh	N-12	A
Lakecreek	P-8	1954
Lake O' Woods	P-10	1955
Langlois	N-1	1954
Leaburg	I-8	1951
Lebanon	G-7	C1944 B
Long Creek	G-22	1951
Lookout Mountain	H-17	1951
Lowell	J-7	1955
Lucile	C-33	B
Lyons	F-8	1951
Mace Mountain	L-7	1955
Madras 1	F-14	B
Madras 2	F-13	B
Madras 3	G-13	B
Madras 4	G-14	B
Mapleton	I-3	1945 A
Marcola	I-7	1952
Marial	O-3	1954
Marys Peak	G-4	1942 A
McKenzie Bridge	I-10	A
McMinnville	E-6	1943
Medford	P-7	1954
Mill City	F-9	1955
Mineral	G-30	A
Molalla	E-8	C1943
Monroe	H-5	C1942 A
Monument	F-21	1951
Moores Hollow	I-30	1951
Mount Angel	E-7	1943
Mount Emily	Q-2	1954
Mount Hood 1	D-12	B
Mount Hood 2	D-11	B
Mount Jefferson 1	F-12	B
Mount Jefferson 2	F-11	B
Mount Jefferson 3	G-11	B
Mount Jefferson 4	G-12	B
Mount McLoughlin	P-9	1955
Mount Vernon	H-22	1943
Mount Wilson	E-12	1956
Nehalem	C-3	1955
Oakridge	K-9	1956
Ochoco Reservoir	H-16	1950
Olds Ferry	H-30	1952
Oregon Caves	Q-5	1954
Oregon City	D-8	1945 A
Pearsoll Peak	P-3	1954
Pelican Butte	O-10	1955
Picture Gorge	G-20	1955
Portland	C-8	A
Port Orford	O-1	1954
Post	I-17	1951
Powers	N-2	1954
Prospect	N-9	1956
Prospect 1	N-10	B
Quartz Mountain	M-8	1955
Quartzville	G-9	1956
Red Butte	M-7	1955
Reedsport	K-2	1956
Richmond	G-19	1953
Riley 1	J-20	B
Ritter	F-22	1952
Roman Nose Mount.	J-4	1945
Roseburg	M-5	1955
Ruch	Q-6	1954
Rustler Peak	O-9	1955
Saddle Mountain	B-4	1955
Salem	F-6	1940 A
Sardine Butte	J-9	A
Scottsburg	K-3	1955
Selma	P-4	1954
Sheridan	E-5	C1942 A
Siltcoos Lake	J-2	1942 A
Sitkum	M-3	1955
Sled Springs	C-29	A
Snow Peak	G-8	1951
Sparta	F-29	A
Spray	F-19	1953
Stayton	F-7	C1944 A
St. Helens	B-7	1954
Summit Lake	L-10	1956
Surveyor Mountain	Q-10	1955
Susanville	G-23	1951
Sutherlin	L-5	1954
Svensen	A-4	1955
Swan Lake	P-12	A
Sweet Home	H-8	1951
Sycan Marsh 1	N-14	B
Sycan Marsh 2	N-13	B
Sycan Marsh 3	O-13	B
Sycan Marsh 4	O-14	B
Talent	Q-7	1954
The Dalles	C-14	A
Three Sisters 1	H-12	B
Three Sisters 2	H-11	B
Three Sisters 3	I-11	B
Three Sisters 4	I-12	B
Tidewater	H-3	1945
Tillamook	D-3	1955
Tiller	N-7	1946
Timber	C-5	1955
Toketee Falls	L-9	1956
Toledo	G-3	1946 A
Trail	O-7	1945
Tualatin	D-7	1943
Tyee	L-4	1955
Valsetz	F-4	C1942 A
Vernonia	B-6	1955
Vistillas 2	P-15	B
Vistillas 3	Q-15	B
Waldo Lake	K-10	A
Waldport	H-2	C1942 A
Wasco	C-16	A
White Salmon	C-13	A
Wimer	O-6	1954
Wishram	C-15	A
Yamhill	D-6	1942
Yaquina	G-2	1946 A

7½-Minute Quadrangles
(Scale 1:24,000)

Name	Location	Date
Amity	49	A
Astoria	3	C1949
Ballston	48	1956
Beaverton	27	1954
Camas	22	1954
Canby	36	1954
Carlton	32	1957
Cathlamet Bay	4	C1949
Clatsop Spit	1	1951
Colton	46	1955
Corbett	23	A
Damascus	30	1954
Dayton	41	A
Deer Island	12	1954
Dixie Mountain	14	1956
Dundee	33	1956
Estacada	39	1954
Forest Grove	17	1956
Gaston	24	1956
Gearhart	5	C1949
Gervais	51	A
Gladstone	29	1954
Green Mountain	7	C1949
Hillsboro	18	1954
Kalama	11	1953
Kelso	9	1953
Laurelwood	25	1956
Linnton	19	1954
Malheur Butte	60	1951
McMinnville	40	A
Mission Bottom	50	1957
Molalla	45	1954
Moores Hollow	58	1951
Mt. Tabor	21	1954
Newberg	34	1954
Olds Ferry	56	1952
Olds Ferry NW	55	1952
Olds Ferry SE	57	1952
Olney	6	C1949
Oregon City	37	1954
Oswego	28	1954
Payette	61	1951
Portland	20	1954
Rainier	10	1953
Redland	38	1954
Sandy	31	1954
Sauvie Island	15	1954
Scholls	26	1954
Scotts Mills	53	1954
Sheridan	47	1956
Sherwood	35	1954
Silverton	52	1956
St. Helens	13	1954
St. Paul	42	1956
Tillamook Head	8	C1949
Vancouver	16	1954
Warrenton	2	C1953
Weiser South	59	1951
Wilhoit	54	1955
Woodburn	43	A
Yoder	44	1955

Special Map

	Location	Date
Crater Lake National Park	N-10	1:62,000

30-Minute Quadrangles
(Scale 1:125,000)

Name	Location	Date
Arlington	C-18	1941
Baker	G-28	1934
Bend	I-14	1940
Blalock Island	C-20	1944
Chemult	M-12	1941
Condon	E-18	1916
Dayville	I-20	1936
Dufur	E-14	1945
Hood River	C-12	1940
Ironside Mountain	I-26	1908
Madras	G-14	1931
Maiden Peak	K-12	1944
Mitchell	G-18	1926
Mitchell Butte	K-30	1921
Mount Hood	E-12	1944
Mount Jefferson	G-12	1938
Newberry Crater	K-14	1935
Pendleton	C-24	1935
Sumpter	G-26	1939
Telocaset	E-28	1932
The Dalles	C-14	1941
Three Sisters	I-12	1941
Umatilla	C-22	1921

KEY TO SYMBOLS

A — Map scheduled for publication 7-1-58 to 6-30-59

B — Map scheduled for publication 7-1-59 to 6-30-60

C — Map published by U. S. Army 29th Engineers

STATE DEPARTMENT OF GEOLOGY
AND MINERAL INDUSTRIES
1069 STATE OFFICE BUILDING
PORTLAND 1 OREGON

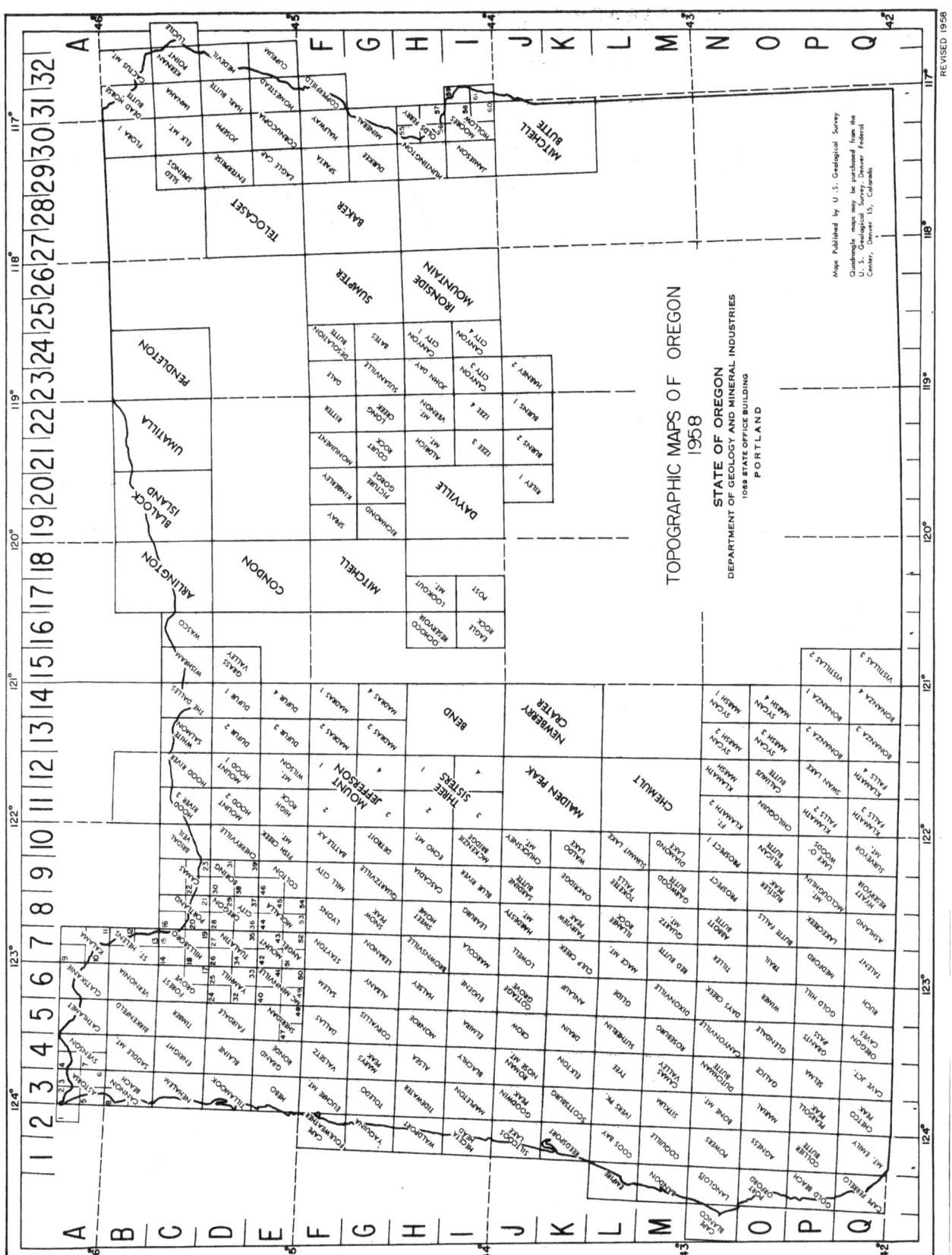

TOPOGRAPHIC MAPS OF OREGON
1958

STATE OF OREGON
DEPARTMENT OF GEOLOGY AND MINERAL INDUSTRIES
1069 STATE OFFICE BUILDING
PORTLAND

Maps Published by U. S. Geological Survey

Quadrangle maps may be purchased from the
U. S. Geological Survey, Denver Federal
Center, Denver 15, Colorado

REVISED 1958

TOPOGRAPHIC MAPS OF OREGON—1948

15-Minute Quadrangles
(Scale 1:62,500)

	Name	Location coordinates		Contour interval	Date last printed
*	Abbott Butte	O	8	50	1947
*	Albany	H	6	25	1944
*	Aldrich Mountain	I	21	50	1943
o	Alsea	I	4	50	1942
o	Astoria	B	3	20	1939
*	Bandon	N	1	50	1944
*	Bates	H	24		Underway
o	Blachly	J	4	50	1942
**	Boring	E	9	25	1944
o	Bridal Veil	D	10	100	1942
x	Brownsville	I	7	25	1938
*	Camas	D	9	25	1942
*	Cape Falcon	C	3	50	1940
*	Cape Foulweather	G	2	50	1944
o	Cathlamet	B	5	20	1941
o	Clatskanie	B	6	25	1942
*	Coos Bay	M	2	50	1945
o	Coquille	N	2	50	1945
o	Corvallis	H	2	50	1942
o	Cottage Grove	K	6	25	1921
*	Crow	K	5	50	1945
o	Dallas	G	5	50	1942
*	Dutchman Butte	O	4		Underway
*	Eagle Rock	I	16	50	1948
*	Elmira	J	5	50	1942
*	Empire	M	1	50	1944
o	Enright	D	4	100	1941
o	Euchre Mountain	G	3	50	1943
**	Eugene	J	6	5 & 10	1940
o	Fairdale	E	5	100	1942
o	Gales Creek	D	6	25	1943
*	Galice	P	4	50	1948
o	Ginger Peak	E	4	100	1942
o	Goodwin Peak	K	3	50	1943
**	Halsey	I	6	10 & 25	1941
**	Heceta Head	J	2	50	1944
x	He Devil	E	32	50	1922
"	Hillsboro	D	7	25	1943
"	John Day	I	23	50	1943
o	Kalama	B	7	20	1943
o	Keasey	C	5	100	1943
**	Lebanon	H	7	25	1944
*	Long Creek	H	22		Underway
*	Mapleton	J	3	50	1945
*	Marys Peak	H	4	50	1942
*	McMinnville	F	6	25	1943
*	Molalla	F	8	25	1943
o	Monroe	I	5	50	1942
*	Mount Angel	F	7	25	1943
*	Mount Vernon	I	22	50	1943
o	Nehalem	D	3	100	1943
o	Nestucca Bay	F	3	100	1942
*	Ochoco Reservoir	I	16		Underway
o	Oregon City	E	8	25	1945
*	Portland	D	8	25	1940
*	Post	J	17		Underway
**	Reedsport	L	2	50	1942
*	Roman Nose Mountain	K	4	100	1944
o	Saddle Mountain	C	4	100	1945
**	Salem	G	6	25	1940
x	Seven Devils	F	32	50	1920
o	Sheridan	F	5	100	1942
**	Siltcoos Lake	K	2	50	1942
o	Spirit Mountain	F	4	100	1942
*	Stayton	G	7	25	1944
o	St. Helens	C	7	25	1943
*	Susanville	H	23		Underway
o	Svensen	B	4	20	1940
*	Tidewater	I	3	50	1945
o	Tillamook	E	3	100	1942
*	Tiller	O	7	50	1946
o	Timber	D	5	100	1941
*	Toledo	H	3	50	1946
*	Trail	P	7	50	1945
**	Tualatin	E	7	25	1943
o	Valsetz	G	4	50	1942
o	Vernonia	C	6	25	1943
"	Waldport	I	2	50	1942
o	Yamhill	E	6	100	1942
*	Yaquina	H	2	50	1946

30-Minute Quadrangles
(Scale 1:125,000)

	Name	Location coordinates		Contour interval	Date last printed
*	Arlington	CD	17-18	50	1941
*	Baker	GH	27-28	100	1934
*	Bend	IJ	13-14	50	1940
*	Blalock Island	CD	19-20	50	1944
*	Chemult	MN	11-12	50	1941
*	Condon	EF	17-18	50	1916
*	Coos Bay	MN	1-2	100	1937
*	Dayville	IJ	19-20	100	1936
*	Diamond Lake	MN	9-10	100	1926
*	Dufur	EF	13-14	100	1945
*	Estacada	EF	9-10	100	1938
*	Grants Pass	QR	5-6	100	1930
*	Hood River	CD	11-12	100	1940
*	Ironside Mountain	IJ	25-26	100	1908
*	Kirby	QR	3-4	100	1942
*	Lowell	KL	7-8	100	1942
*	Madras	GH	13-14	100	1931
*	Maiden Peak	KL	11-12	100	1944
*	McKenzie Bridge	IJ	9-10	100	1940
*	Medford	QR	7-8	100	1945
*	Mill City	GH	9-10	100	1941
*	Mitchell	GH	17-18	100	1926
*	Mitchell Butte	KL	29-30	50	1921
*	Mt. Hood	EF	11-12	100	1944
*	Mt. Jefferson	GH	11-12	100	1938
*	Newberry Crater	KL	13-14	100	1935
*	Pendleton	CD	23-24	50	1935
*	Pine	GH	29-30	100	1941
*	Port Orford	OP	1-2	100	1944
*	Riddle	OP	5-6	100	1942
*	Roseburg	MN	5-6	100	1942
*	Sumpter	GH	25-26	100	1939
*	Telocaset	EF	27-28	100	1932
*	The Dalles	CD	13-14	50	1941
*	Three Sisters	IJ	11-12	100	1941
*	Umatilla	CD	21-22	50	1921
*	Waldo Lake	KL	9-10	100	1944
*	Weiser	IJ	31-32	100	1916

SPECIAL MAPS

Name	Location coordinates		Contour interval	Scale
Crater Lake Nat'l Park	NO	9-10	50	1:62,000
Crater Lake and vicinity	NP	9-11	50	1:48,000
Mt. Hood and vicinity	DE	10-11	100	1:125,000
Picture Gorge (advance)	HI	20	5	1:24,000
Squaw Butte Ranch (advance)	M	19-20	50	1:48,000

River Surveys
(Various scales and contour intervals)

Name	Location coordinates		Date last printed
Applegate River, 4 sheets	QR	5-6	1940
Catherine Creek, 1 sheet	F	28	1933
Chewaucan, 3 sheets	PQ	16	1938
Clackamas River (W.S.P. 349)	F	10	1914
Columbia River	B-D	3-22	1929-30
Coquille River	N	1-2	1926
Cow Creek, 1 sheet	O	6	1939
Crooked River	I	14-15	1926
Deep Creek and Camas Creek, 4 sheets	R	18-19	1939
Deschutes River (W.S.P. 344)	D-J	14-15	1911
Evans Creek, 2 sheets	PQ	6	1938
Gales Creek, 1 sheet	D	5	1934
Grande Ronde River, 7 sheets	CD	27-29	1937
Grave Creek, 1 sheet	P	6	1938
Hood River, 4 sheets	DE	12	1939
Illinois River (see Rogue River)			
John Day River (W.S.P. 377)	D-G	16-20	1909
Jump-off Joe Creek, 1 sheet	P	5	1937
Klamath River, 16 sheets	R-T	5-10	1926
Little Butte Creek, 3 sheets	Q	7-8	1938
Lookout Point, 4 sheets	K	9	1938
Luckiamute River, 1 sheet	GH	4-5	1938
McKenzie River, 6 sheets	J	6-10	1926
Metolius River	H	12-13	1912
Nehalem River, 7 sheets	C	4-6	1938
North Santiam River, 2 sheets	GH	8-11	1944
North Umpqua River	M	5-10	1923
Rogue River, 14 sheets	O-Q	1-9	1925
Sandy River (W.S.P. 348)	E	11	1927
Santiam River (W.S.P. 349)	GH	8-11	1914
Santiam River (see North and South Santiam)			
Separation Creek	J	10	1928
Siletz River	GH	3	1926
Snake River, 17 sheets	C-K	30-33	1939
South Santiam River, 5 sheets	I	8-9	1938
South Umpqua River, 3 sheets	O	6-7	1938
South Yamhill River, 2 sheets	F	4	1938
Umatilla River, 3 sheets	D	25	1938
Umpqua River, 9 sheets	LM	3-4	1926
Walla Walla River, 4 sheets	C	25-26	1932
White River, 3 sheets	F	12-14	1932
Willamette River (W.S.P. 349 and 378)	KL	9-10	1938
Willamette River (see Lookout Point)			
Willamina Creek, 1 sheet	F	5	1937
Yamhill River (see South Yamhill)			

KEY TO SYMBOLS

*   Map published by U.S. Geological Survey, obtainable from the Director, U.S. Geol. Survey, Washington, D.C.

x   Map published by U.S. Geological Survey, quadrangle incomplete.

o   Map printed by the U.S. Army, 29th Engineers.

*   Map revised by U.S. Army, 29th Engineers, on U.S. Geol. Survey topographic base.

STATE DEPARTMENT OF GEOLOGY AND MINERAL INDUSTRIES

702 WOODLARK BUILDING
PORTLAND 5, OREGON

CARRYING SACK

Some sort of packsack for tools and equipment and for bringing out specimens, is necessary. Some persons prefer a packsack of the Trapper Nelson, or rucksack, type; others, a knapsack with a single shoulder strap. The packsack should be stout and serviceable, waterproof, good-sized, and comfortable to the wearer.

NOTE BOOK AND PENCIL

A note book and pencil are essential to one who wishes to jot down observations. Data should be recorded as to the exact locality, map used, township, range, section, sub-division of section, and, if possible, a description of the locality, and a field identification of the rock or specimen. The locality should be given some sort of distinguishing number.

Many collectors prefer a loose-leaf note book of pocket size. A six-ring binder is preferable to a three-ring, as it tends to hold the paper more satisfactorily. Note books bound with a spiral binder are very good. The loose sheets may then be torn out to serve as a label to accompany the specimen or sample.

A specimen label is illustrated in figure 12. This label has been extensively used and found satisfactory.

Figure 12.   Specimen Field Label.

LOCATION: County . . . . Date . . . .
        Quadrangle . . . . .Elevation. . . . . Sample No. . . .
             Section_____, T._____, R._____

                                                  Notebook
                                                  Page . . . .

FIELD DESCRIPTION:

Property Owner . . . . Collector . . . .

WRAPPING PAPER.

It is advisable to use paper for wrapping specimens. Some collectors prefer paper sacks of the "nail-sack" variety in "four-pound" and "twelve-pound" sizes. The specimen is placed in the sack with the label, and the extra paper

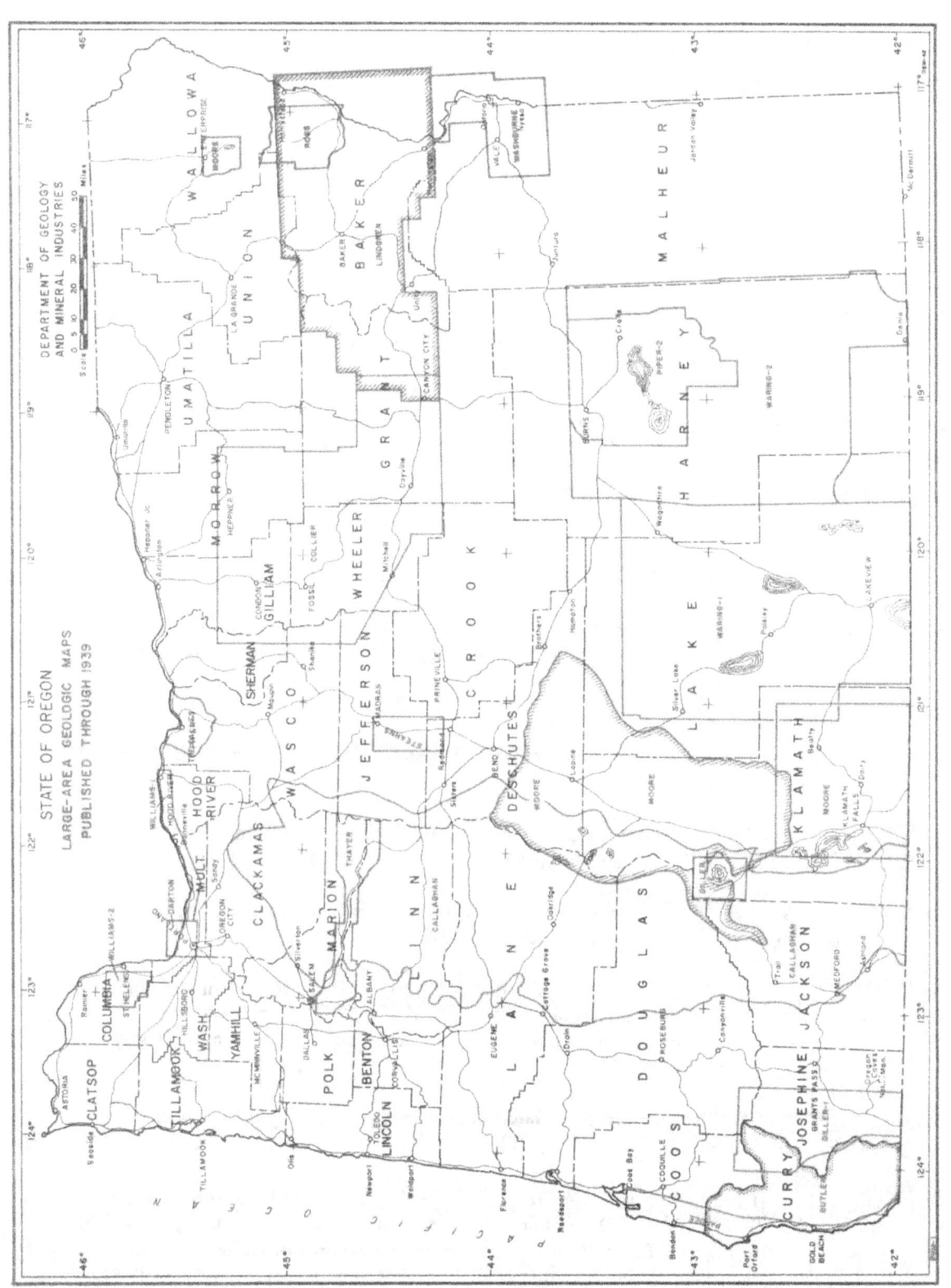

STATE OF OREGON
LARGE-AREA GEOLOGIC MAPS
PUBLISHED THROUGH 1939

DEPARTMENT OF GEOLOGY
AND MINERAL INDUSTRIES

LARGE-AREA GEOLOGIC MAPS PUBLISHED THROUGH 1939

Butler, G.M. (and Mitchell, G.J.), Preliminary survey of the geology and mineral resources of Curry County, Oregon: Oreg. Bur. Mines & Geol. Min. Res. of Oregon, vol.2, no.2, 1916.

Callaghan, E., (and Buddington, A.F.), Metalliferous mineral deposits of the Cascade Range in Oregon: U.S. Geol. Survey Bull. 893, 1938.

Collier, A.J., The geology and mineral resources of the John Day region: Oreg. Bur. Mines & Geol. Min. Res. of Oregon, vol. 1, no. 3, 1914.

Darton, N.H., Structural materials in parts of Oregon and Washington: U.S. Geol. Survey Bull. 387, 1909.

Diller, J.S., (1) Mineral resources of southwestern Oregon: U.S. Geol. Survey Bull. 546, 1914; (2) (and Patton, H.B.), The geology and petrography of Crater Lake National Park: U.S. Geol. Survey Prof. Paper 3, 1902.

Lindgren, W., The gold belt of the Blue Mountains of Oregon: U.S. Geol. Survey 22nd Ann. Rpt., pt. 2, p. 551-776, 1901.

Moore, B.N., Nonmetallic mineral resources of eastern Oregon: U.S. Geol. Survey Bull. 875, 1937.

Pardee, J.T., Beach placers of the Oregon coast: U.S. Geol. Survey Circ. 8, 1934.

Piper, A.M., (1) Geology and ground-water resources of The Dalles region, Oregon: U.S. Geol. Survey Water-Supply Paper 659-B, 1932; (2) (and Robinson, T.W., and Park, C.F.), Geology and ground-water resources of the Harney basin, Oregon: U.S.Geol. Survey Water-Supply Paper 841, 1939.

Ross, C.P., (1) Geology of part of the Wallowa Mountains, Oregon: Oreg. Dept. Geol. & Min. Ind. Bull. 3, 1938.

Stearns, H.T., Geology and water resources of the Middle Deschutes River basin, Oregon: U.S. Geol. Survey Water-Supply Paper 637-D, 1931.

Thayer, T.P., Geology of the Salem Hills and North Santiam River basin, Oregon: Oreg. Dept. Geol. & Min. Ind. Bull. 15, 1939.

Waring, C.A., (1) Geology and water resources of a portion of south-central Oregon: U. S. Geol. Survey Water-Supply Paper 220, 1908; (2) Geology and water resources of the Harney basin region, Oregon: U.S.Geol. Survey Water-Supply Paper 231, 1909.

Washburne, C.W., Gas and oil prospects near Vale, Oregon, and Payette, Idaho: U.S. Geol. Survey Bull. 431-A, 1911.

Williams, I.A., (1) The Columbia River Gorge - its geologic history interpreted from the Columbia River Highway: Oreg. Bur. Mines & Geol. Min. Res. of Oregon, vol.2, no. 3, 1916; (2) (and Parks, H.M.), The limonite iron ores of Columbia County, Oregon: Oreg. Bur. Mines & Geol. Min. Res. of Oregon, vol. 3, no. 3, 1923.

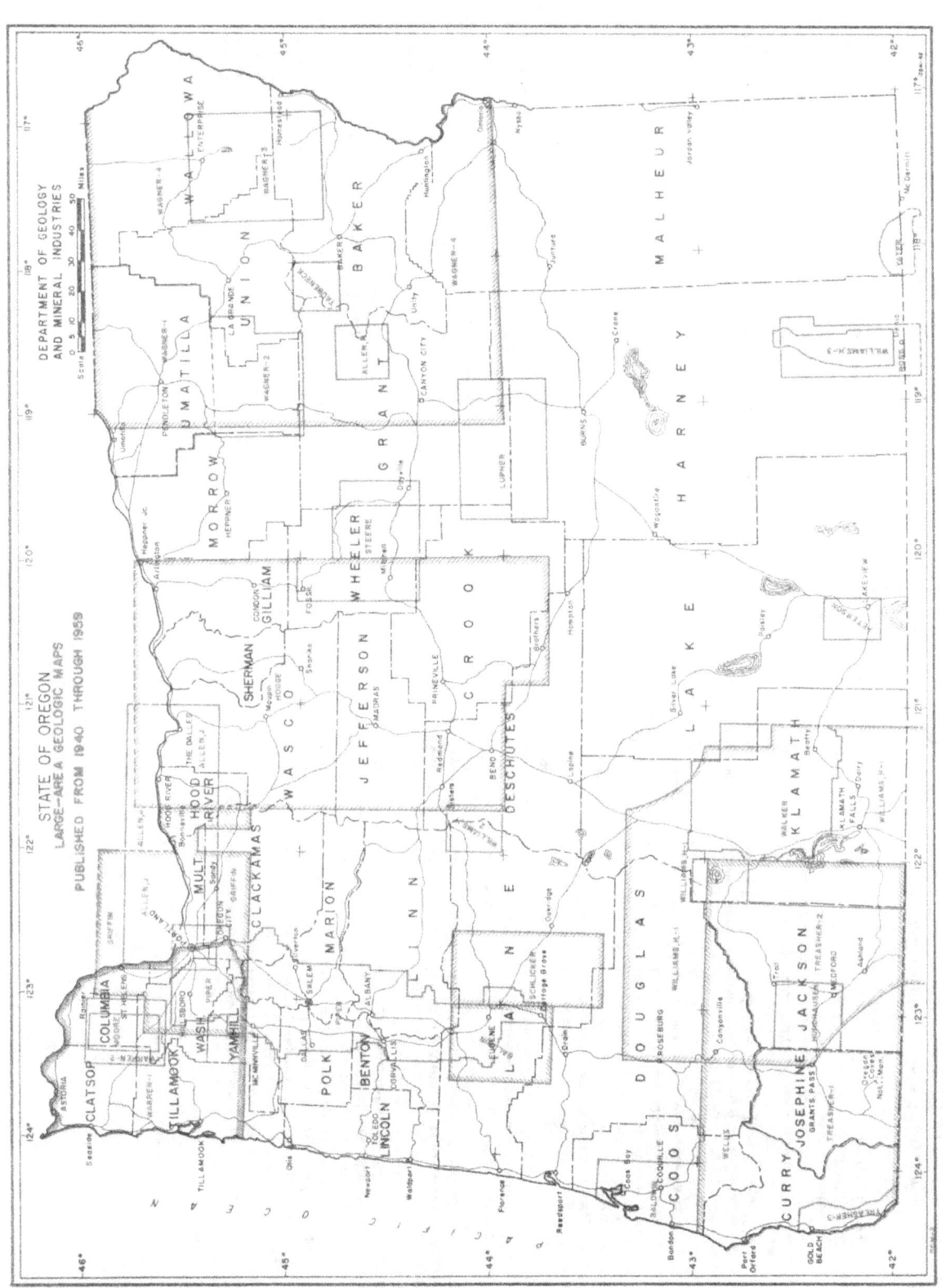

STATE OF OREGON
LARGE-AREA GEOLOGIC MAPS
PUBLISHED FROM 1940 THROUGH 1959

DEPARTMENT OF GEOLOGY
AND MINERAL INDUSTRIES

Scale
0 5 10 20 30 40 50 Miles

Allen, J.E., Columbia River Gorge: Guidebook for Geol. Soc. Am. field trip excursions: Univ. of Oregon, 1958; Field guidebook: Oreg. Dept. Geol. & Min. Ind. Bull. 50, 1959.

Allen, R.M., Jr., Geology and mineralization of the Morning mine and adjacent region, Grant County, Oregon: Oreg. Dept. Geol. & Min. Ind. Bull. 39, 1948.

Baldwin, E.M., Field trip, Eugene to Coos Bay via Reedsport (in: Field Guidebook): Oreg. Dept. Geol. & Min. Ind. Bull. 50, 1959.

Griffin, W.C., (et al), Water resources of the Portland, Oregon, and Vancouver, Washington, area: U.S.Geol. Survey Circ. 372, 1956.

Hodge, E.T., Geologic map of north central Oregon: Oreg. State College Mon., Studies in Geol., no.3, 1942.

Hundhausen, R.H., Investigation of Shamrock copper-nickel mine, Jackson County, Oregon: U.S. Bur. Mines Rpt. Invest. 4895, 1952.

Lupher, R.L., Jurassic stratigraphy of central Oregon: Geol. Soc. America Bull., vol. 52, no. 2, p.219-269, 1941.

Moore, R.C., and Vokes, H.E., Lower Tertiary crinoids from northwestern Oregon: U.S. Geol. Survey Prof. Paper 233-E, 1953.

Peterson, N.V., Preliminary geology of the Lakeview uranium area, Oregon: Oreg. Dept. Geol. & Min. Ind. Ore.-Bin, vol. 21, no. 2, 1959.

Piper, A.M., Ground-water resources of the Willamette Valley, Oregon: U.S. Geol. Survey Water-Supply Paper 890, 1942.

Ross, C.P., Quicksilver deposits in the Steens and Pueblo mountains, southern Oregon: U.S. Geol. Survey Bull. 931-J, 1942.

Schlicker, H.G., and Dole, H.M., Reconnaissance geology of the Marcola, Leaburg, and Lowell quadrangles, Oregon: Oreg. Dept. Geol. & Min. Ind. Ore.-Bin, vol. 19, no. 7, p.57-62, 1957.

Steere, M.L., Geology of the John Day country, Oregon: Oreg. Dept. Geol. & Min. Ind. Ore.-Bin, vol. 16, no. 7, 1954.

Taubeneck, W.H., Geology of the Elkhorn Mountains, northeastern Oregon; Bald Mountain batholith: Geol. Soc. Am. Bull., vol. 68, p.181-238, 1957.

Treasher, R.C., (1) Geologic map of Josephine County (Oregon Metal Mines Handbook): Oreg. Dept. Geol. & Min. Ind. Bull. 14-C, vol. 2, sec. 1, 1942; (2) Geologic map of Jackson County (Oregon Metal Mines Handbook): Oreg. Dept. Geol. & Min. Ind. Bull.14-C, vol. 2, sec. 2, 1943; (3) Reconnaissance geologic survey in Curry County along Coast Highway from Gold Beach to California State line: Geol. Soc. Oreg. Country Geological News Letter, vol. 9, no. 13, 1943.

Wagner, N.S. (1) Ground-water studies in Umatilla and Morrow counties, Oregon: Oreg. Dept. Geol. & Min. Ind. Bull. 41, 1949; (2) Geology of the southern half of Umatilla County, Oregon: Oreg. Dept. Geol. & Min. Ind. Ore.-Bin, vol. 16, no.3, 1954; (3) Summary of Wallowa Mountains geology, Oregon: Oreg. Dept. Geol. & Min. Ind. Ore.-Bin, vol. 17, no.5, 1955; (4) Important rock units of northeastern Oregon: Oreg. Dept. Geol. & Min. Ind. Ore.-Bin, vol. 20, no. 7, p.63-68, 1958.

Walker, G.W., Pumice deposits of the Klamath Indian Reservation, Klamath County, Oregon: U.S. Geol. Survey Circ. 128, 1951.

Warren, W.C., (1) (et al), Geology of northwestern Oregon, west of the Willamette River and north of latitude 45°15': U.S. Geol. Survey Map OM 42, 1945; (2) (and Norbisrath, H.,) Stratigraphy of upper Nehalem River basin, north-western Oregon: Am. Assoc. Petroleum Geol. Bull., vol.30, no. 2, 1946.

Wells, F.G., Preliminary geologic map of southwestern Oregon west of meridian 122° west, and south of parallel 43° north: U.S. Geol Survey Map MF 38, 1955.

Williams, H., (1) The geology of Crater Lake National Park, Oregon, with a recon-naissance of the Cascade Range southward to Mount Shasta: Carnegie Inst. Wash. Pub. 540, 1942; (2) Volcanoes of the Three Sisters region, Oregon Cascades: California Univ. Dept. Geol. Sci. Bull., vol.27, no. 3, 1944; (3) (and Compton, R.R.) Quicksilver deposits of Steens Mountain and Pueblo Mountains, southeast Oregon: U.S. Geol. Survey Bull. 995-B, 1953.

Yates, R.G., Quicksilver deposits of the Opalite district, Malheur County, Oregon, and Humboldt County, Nevada: U.S. Geol. Survey Bull. 931-N, 1942.

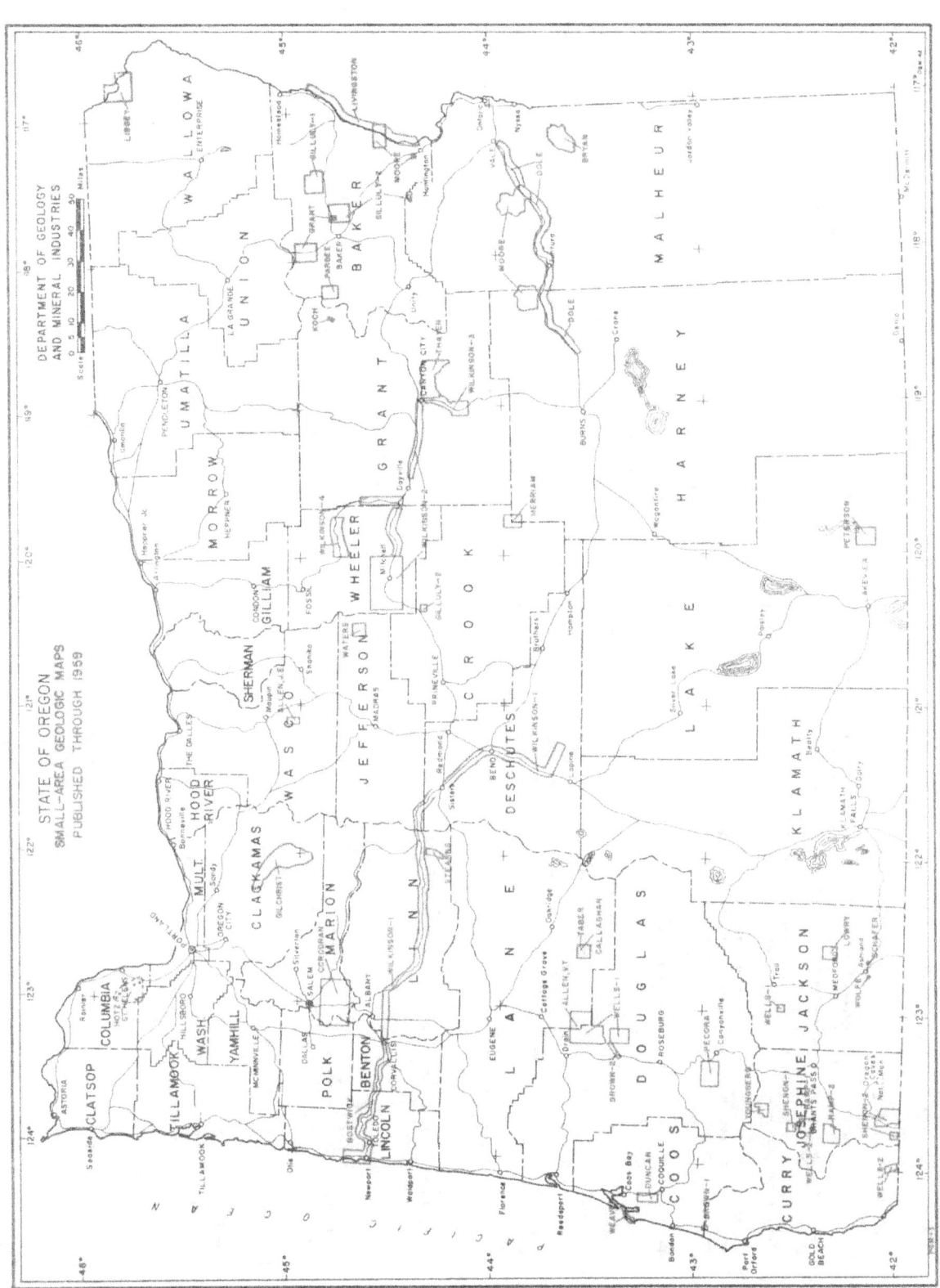

STATE OF OREGON
SMALL-AREA GEOLOGIC MAPS
PUBLISHED THROUGH 1959

DEPARTMENT OF GEOLOGY
AND MINERAL INDUSTRIES

Scale
0 5 10 20 30 40 50 Miles

Allen, J.E., Perlite deposits near the Deschutes River, southern Wasco County, Oregon: Oreg. Dept. Geol. & Min. Ind. Short Paper 16, 1946.

Allen, V.T., Loofbourow, J.S., and Nichols, R.L., The Hobart Butte high-alumina clay deposit, Lane County, Oregon: U.S. Geol. Survey Circ. 143, 1951.

Bostwick, D.A., Field trip, Corvallis to Depoe Bay via Newport (in: Field Guidebook): Oreg. Dept. & Min. Ind. Bull. 50, 1959.

Brown, R.E., (1) Some manganese deposits in the southern Oregon coastal region: Oreg. Dept. Geol. & Min. Ind. Short Paper 9, 1942, (2) (and Waters, A.C.), Quicksilver deposits of the Bonanza-Nonpareil district, Douglas County, Oregon: U.S. Geol. Survey Bull. 955-F, 1951.

Bryan, K., Geology of the Owyhee irrigation project:U.S.Geol.Survey Water-Supply Paper 597-A, 1929.

Callaghan, E., (and Buddington, A.F.) Metalliferous mineral deposits of the Cascade Range in Oregon (Geol. map of Bohemia dist.): U.S. Geol. Survey Bull. 893, 1938.

Corcoran, R.E., (and Libbey, F.W.), Ferruginous bauxite deposits in the Salem Hills, Marion County, Oregon: Oreg. Dept. Geol. & Min. Ind. Ore.-Bin, vol.17, no.4,1955; Oreg. Dept. Geol. & Min. Ind. Bull. 46, 1956.

Dole, H.M., (and Corcoran, R.E.), Reconnaissance geology along U.S.Highway 20 between Vale and Buchanan, Malheur and Harney counties, Oregon: Oreg. Dept. Geol. & Min. Ind. Ore.-Bin, vol. 16, no. 6, 1954.

Duncan, D.C., Geology and coal deposits in part of the Coos Bay coal field, Oregon: U.S. Geol. Survey Bull. 982-B, 1953.

Gilchrist,F.G.,Clackamas River field trip:Geol.Soc.Oregon Country Geol.News Letter,vol.18,no.9,1952.

Gilluly, J.,(1)Copper deposits near Keating, Oregon:U.S.Geol.Survey Bull.830-A,1933;(2)(and Reed,J.C. and Park,C.F.), Some mining districts of eastern Oregon:U.S.Geol.Survey Bull. 846-A, 1933.

Grant, U.S.,(and Cady, G.H.), Preliminary report on the general and economic geology of the Baker district of eastern Oregon: Oreg. Bur.Mines & Geol.Min.Res.of Oreg.,vol.1, no.6, 1914.

Hotz, P.E., Limonite deposits near Scappoose, Columbia County, Oregon: U.S.Geol. Survey Bull. 982-C, 1953.

Koch, G.S., Jr., Lode mines of the central part of the Granite mining district, Oregon: Oreg. Dept. Geol. & Min. Ind. Bull. 49, 1959.

Libbey, F.W., Some mineral deposits in the area surrounding the junction of the Snake and Imnaha rivers in Oregon: Oreg. Dept. Geol. & Min. Ind. Short Paper 11, 1943.

Livingston, D.C., A geologic reconnaissance of the Mineral and Cuddy Mountain mining districts, Washington and Adams counties, Idaho (adjacent to Snake River): Idaho Bur. Mines & Geol. Pamphlet 13, 1925.

Lowry, W.D., Tyrrell manganese deposit and other similar properties in the Lake Creek district, Oregon: Oreg. Dept. Geol. & Min. Ind. Short Paper 10, 1943.

Merriam, C.W., (and Berthiaume, S.A.), Late Paleozoic formations in central Oregon: Geol. Soc. Am. Bull., vol.54, no.2, p.145-171, 1943.

Moore, B.N., Nonmetallic mineral resources of eastern Oregon:U.S.Geol.Survey Bull. 875, 1937.

Pardee, J.T., Faulting and vein structure in the Cracker Creek gold district, Baker County, Oregon: U.S. Geol. Survey Bull. 380, p.85-93, 1909.

Pecora, W.T., (and Hobbs, W.S.), Nickel deposits near Riddle, Douglas County, Oregon: U.S. Geol. Survey Bull. 931-I, 1942.

Peterson, N.V., Lake County's new continuous geyser: Oreg. Dept.Geol.& Min. Ind. Ore.-Bin, vol. 21, no. 9, 1959.

Ramp, L., (1) Geologic map of Chrome Ridge area, Josephine County, Oregon: Oreg. Dept. Geol. & Min. Ind. Ore.-Bin, vol. 18, no.3, 1956; (2) Geology of the lower Illinois River chromite district: Oreg. Dept. Geol. & Min. Ind. Ore.-Bin, vol. 19, no. 4, p.29-34, 1957.

Schafer, Max, Occurrence and utilization of carbon-dioxide-rich water near Ashland, Oregon: Oreg. Dept. Geol. & Min. Ind. Ore.-Bin, vol. 17, no. 7, 1955.

Shenon, P.J., (1) Geology of the Robertson, Hundinger, and Robert E. gold mines, southwestern Oregon: U.S.Geol. Survey Bull. 830-B, 1923; (2) Geology and ore deposits of the Takilma-Waldo district, Oregon: U.S.Geol. Survey Bull. 846-B, 1933.

Stearns, H.T., Geology and water resources of the upper McKenzie River Valley, Oregon: U.S.Geol. Survey Water-Supply Paper 597-D, 1929.

Taber, J.W., A reconnaissance of lode mines and prospects in the Bohemia mining district, Lane and Douglas counties, Oregon: U.S. Bur.Mines Inf.Circ.7512, 1949.

Thayer, T.P., Chromite deposits of Grant County, Oregon: U.S.Geol.Survey Bull. 922-D, 1940.

Waters, A.C., (et al), Quicksilver deposits of the Horse Heaven mining district, Oregon: U.S.Geol. Survey Bull. 969-E, 1951.

Weaver, C.E., Stratigraphy and paleontology of the Tertiary formations at Coos Bay, Oregon: Washington Univ. Pub. in Geol., vol. 6, no. 2, 1945.

Wells,F.G.,(1)(and Waters,A.C.), Quicksilver deposits of southwestern Oregon: U.S.Geol.Survey Bull. 850, 1934;(2)(and Page,L.R.,and James,H.L.), Chromite deposits in the Sourdough area, Curry County, and the Briggs Creek area, Josephine County, Oregon: U.S.Geol.Survey Bull.922-P, 1940.

Wilkinson, W.D. (1) (and Schlicker, H.G.), Field trip, Corvallis to Prineville via Bend and Newberry Crater (in: Field Guidebook): Oreg.Dept.Geol.& Min.Ind.Bull. 50, 1959; (2) Field trip, Prineville to John Day via Mitchell (in: Field Guidebook): Oreg.Dept.Geol.& Min.Ind.Bull.50, 1959; (3) (and Thayer,T.P.), Field trip, John Day to upper Bear Valley (in: Field Guidebook): Oreg.Dept.Geol.& Min.Ind.Bull.50, 1959; (4) (and Allen,J.E.), Field trip, Picture Gorge to Portland via Arlington (in: Field Guidebook): Oreg.Dept.Geol.& Min.Ind.Bull.50, 1959.

Wolfe, H.D.,(and White, D.J.), Preliminary report on tungsten in Oregon: Oreg.Dept.Geol.& Min. Ind. Short Paper 22, 1951.

Youngberg, E.A., Mines and prospects of the Mount Reuben mining district, Josephine County, Oregon: Oreg.Dept.Geol.& Min.Ind.Bull.34, 1947.

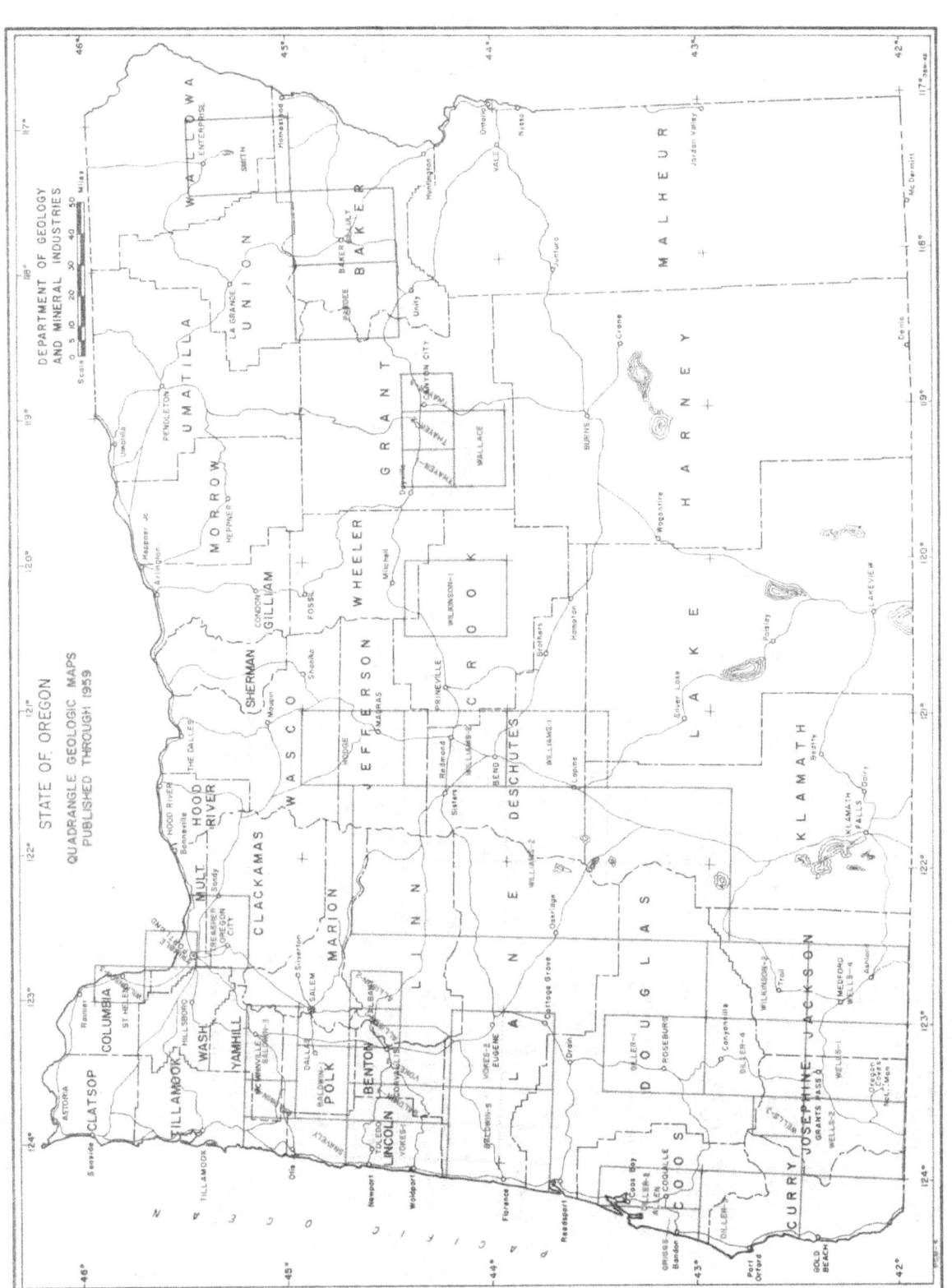

STATE OF OREGON

QUADRANGLE GEOLOGIC MAPS
PUBLISHED THROUGH 1959

DEPARTMENT OF GEOLOGY
AND MINERAL INDUSTRIES

Scale 0 5 10 20 30 40 50 Miles

Allen, J.E. (and Baldwin, E.M.), Geology and coal resources of the Coos Bay quad-rangle, Oregon: Oreg. Dept. Geol. & Min. Ind. Bull. 27, 1944.

Allison, I.S., (1) Geology of the Albany quadrangle, Oregon: Oreg. Dept. Geol. & Min. Ind. Bull. 37, 1953; (2) (and Felts, W.M.), Geology of the Lebanon quadrangle, Oregon: Oreg. Dept. Geol. & Min. Ind. Map with text, 1956.

Baldwin, E.M., (1) Geology of the Dallas and Valsetz quadrangles, Oregon: Oreg. Dept. Geol. & Min. Ind. Bull. 35, 1947; (2) (and Roberts, A.E.) Geology of the Spirit Mountain quadrangle, Oregon: U.S. Geol. Survey Map OM 129, 1952; (3) (et al) Geology of the Sheridan and McMinnville quadrangles, Oregon: U.S. Geol. Survey Map OM 155, 1955; (4) Geology of the Marys Peak and Alsea quadrangles, Oregon: U.S. Geol. Survey Map OM 162, 1955; (5) Geologic map of the lower Siuslaw River area, Oregon: U.S. Geol. Survey Map OM 186, 1956;

Diller, J.S., U.S. Geol. Survey Atlas series: (1) Roseburg folio (no.49), 1898; (2) Coos Bay folio (no.73), 1901; (3) Port Orford folio (no.89), 1903; (4) (and Kay, G.F.), Riddle folio (no.218), 1924.

Gilluly, James, Geology and mineral resources of the Baker quadrangle, Oregon: U.S. Geol. Survey Bull. 879, 1937.

Griggs, A.B., Chromite-bearing sands of the southern part of the coast of Oregon: U.S. Geol. Survey Bull. 945-E, 1945.

Hodge, E.T., Geologic map of the Madras quadrangle, Oregon:Oregon State Coll. Mon., Studies in Geol., 1941.

Pardee, J.T., Preliminary geologic map of the Sumpter quadrangle, Oregon: Oreg. Dept. Geol. & Min. Ind. Map with text, 1941.

Smith, W.D. (and Allen, J.E.), Geology and physiography of the northern Wallowa Mountains, Oregon: Oreg. Dept. Geol. & Min. Ind. Bull. 12, 1941.

Snavely, P.D. (and Vokes, H.E.), The coastal area between Cape Kiwanda and Cape Foulweather, Oregon: U.S. Geol. Survey Map OM 97, 1949.

Thayer, T.P., (1) Preliminary geologic map of the Aldrich Mountain quadrangle, Oregon: U.S. Geol. Survey Map MF 49, 1956;(2) Preliminary geologic map of the Mt. Vernon quadrangle, Oregon: U.S. Geol. Survey Map MF 50, 1956; (3) Pre-liminary geologic map of the John Day quadrangle, Oregon: U.S. Geol. Survey Map MF 51, 1956.

Treasher, R.C., Geologic history and map of the Portland area: Oreg. Dept. Geol. & Min. Ind. Short Paper 7, 1942.

Trimble, D.E., Geology of the Portland quadrangle, Oregon-Wash.: U.S. Geol. Survey Map GQ 104, 1957.

Vokes, H.E. (1) (and Norbisrath, Hans, and Snavely, P.D), Geology of the Newport-Waldport area, Lincoln County, Oregon: U.S. Geol. Survey Map OM 88, 1949; (2) (and Snavely, P.D., and Myers, D.A.), Geology of the southern and southwestern border areas of the Willamette Valley, Oregon: U.S. Geol. Survey Map OM 110, 1951; (3) (and Myers, D.A., and Hoover, Linn), Geology of the west-central border area of the Willamette Valley, Oregon: U.S. Geol. Survey Map OM 150, 1954.

Wallace, R.E. (and Calkins, J.A.), Reconnaissance geologic map of the Izee and Logdell quadrangles, Oregon: U.S. Geol. Survey Map MF 82, 1956.

Wells, F.G. (1) (et al) Preliminary geologic map of the Grants Pass quadrangle, Oregon: Oreg. Dept. Geol. & Min. Ind. Map with text, 1940; (2) (and Hotz, P.E., and Cater, F.W.) Preliminary description of the geology of the Kerby quadrangle, Oregon: Oreg. Dept. Geol. & Min. Ind. Bull. 40, 1949; (3) (and Walker, G.W.) Geology of the Galice quadrangle, Oregon: U.S. Geol. Survey Map GQ 25, 1953; (4) Geology of the Medford quadrangle, Oregon: U.S. Geol. Survey Map GQ 89, 1956.

Wilkinson, W.D., (1) Reconnaissance geologic map of the Round Mountain quadrangle, Oregon: Oreg. Dept. Geol. & Min. Ind. Map with text, 1940; (2) Recon-naissance geologic map of the Butte Falls quadrangle, Oregon: Oreg. Dept. Geol. & Min. Ind. Map, 1941; (3) (and Lowry, W.D., and Baldwin, E.M.), Geology of the St. Helens quadrangle, Oregon: Oreg. Dept. Geol. & Min. Ind. Bull. 31, 1946.

Williams, Howel, (1) Newberry Volcano of central Oregon: Geol. Soc. Am. Bull., vol. 46, p.253-304, 1935; (2) A geologic map of the Bend quadrangle, Oregon, and a reconnaissance geologic map of the central portion of the High Cascade Mountains: Oreg. Dept. Geol. and Min. Ind. Maps with text, 1957.

is folded over and around the specimen.

Wrapping paper, newspaper, or magazine paper the size of the Saturday Evening Post will serve satisfactorily.   A ball of string for tying will prove useful.

MISCELLANEOUS

Matches kept in a waterproofed container may save many an embarrassing moment. If no match safe is available, heat some paraffin until it is liquid, then throw in a handful of wooden matches and allow them to soak in the liquid paraffin for several minutes.   Take them out and stand them upright to drain, the untreated end down.   The match will then stand soaking in water for days and still be usable. Another method used by some woodsmen who wish to protect themselves against setting forest fires with matches is to stand the lower two-thirds of the match stick in water-glass; the water-glass is non-inflammable and fireproofs that end of the match.   The matchhead is then paraffined, care being used that the paraffin covers the margin of the water glass.   When ignited, the match will burn brightly until the flame encounters the water-glassed portion, where it dies.   Another fire-prevention trick is as follows: Carry an empty tobacco can; place all match sticks, tobacco ashes, cigarette butts and pipe heels in the can; and keep can closed.

A small piece of "fat wood", that is, pitchy wood, in the pack is a distinct aid in starting a campfire.   "Kindlesticks", a patented product, are most excellent, and can be purchased for a small sum at most "dime stores".   They burn sufficiently long and with enough heat to ignite damp wood.

It is probable that a great deal of mineral prospecting will be done on public lands.   The U. S. Forest Service officials are usually willing to assist and in return they ask that reasonable precautions be used to prevent setting fire to the forests.   The Forest Officers comprise a body which has been charged with the protection of public property, in which we as taxpayers have an interest.   If one tries to assist them in their work, he will find them a most helpful group.   Give them a fair trial, and ask for an explanation of the various "regulations".

TESTS, GENERAL DESCRIPTION

A mineralogist uses a number of tests which may provide the necessary clues to the identification of a mineral. A well-trained and experienced person may pick up a mineral and tentatively identify it without making tests but one may rest assured that it is a wide experience that enables him to see the answer to many of the tests without actually performing them.

It should be understood that these field tests necessarily do not give a final and complete determination. The collector or prospector frequently may be at a loss to determine a mineral accurately; this should not be discouraging, as there are many minerals - about 1600 - and even experts are often forced to make chemical and microscopic tests. The tests given herein, however, provide means for a tentative determination subject to later check.

There may or may not be a relationship between a mineral name and its composition. . . . . The nomenclature exists and each mineral name must be memorized.

The collector may have difficulty in understanding some of the tests. It can be said that adamantine luster is a very high luster with a peculiar, bright oily flash characteristic of diamond, but this means little if adamantine luster has never been studied. Certain mineral supply houses put out small, low-cost mineral sets to illustrate these features, and many persons may find it desirable to purchase these for study. The principal sets are:

	Approximate Cost
Hardness	$1.50 - $3.50
Luster	2.00 - 12.00
Cleavage	1.75 - 3.00
Tenacity	1.25 - 3.00
Fracture	1.00 - 3.00

Specimens purchased from well-established supply houses may be slightly higher in price but one has more confidence in the accuracy of their identification.

The correct name of a mineral is determined by a process of elimination. Each test is applied in turn, and a great number of possibilities are thus eliminated. For example: if a mineral has a determined hardness of 7, all minerals having a hardness greater or less than 7 are eliminated. Next, the mineral may have a metallic luster, so all minerals of hardness 7 having a non-metallic luster are eliminated. The final result after all tests are applied, is one mineral the name of which is sought. It is seldom that one test alone will identify a mineral, and several, or all, must be applied.

COLOR

Color of a mineral is usually the first feature to be noticed, and may prove the most important. There are certain warnings that should be given, since a given mineral may occur in different colors. Calcite may be colorless, white, blue, pink, amber, or almost any color. On the other hand, epidote has a peculiar pistachio green color that is quite diagnostic.

Most people are aware that the color of a substance seen in artificial light may be quite different from its color in daylight. The color observed is reflected from the surface and all other colors are absorbed. A red article will appear black in a blue light as there are no red rays reflected.

Another difficulty is that individuals differ in their ability to distinguish color. This may range from true color-blindness, in which one color cannot be distinguished from another, to slight differences in observation. One eye may have different color perception than the other.

Therefore, when color is to be used as a test, the following points should be carefully considered:

1. Does this mineral have more than one common color occurrence?
2. Is the light in which the specimen is studied, a true white light such as daylight?
3. Is my color perception standard, and if not just what is the variation?

LUSTER

Luster is the "shine" or "sheen" or "look" of the surface of a mineral under light reflected from its surface.

The two main kinds of luster are metallic and non-metallic.

Metallic Luster

Metallic luster is the peculiar, brilliant shine reflected from metal surfaces and "looks that way" because the mineral is opaque even on the thinnest edges. Galena, pyrite, and many others have this luster. It is important that the amateur become familiar with metallic luster as it is one of the first tests used to separate minerals into different groups.

Non-Metallic Luster

Non-metallic luster is luster other than metallic, and is subdivided into groups characterized by the luster of well-known objects.

Adamantine Luster

Adamantine luster is that bright, somewhat "oily" appearance so characteristic of diamond.

Vitreous luster

Vitreous luster is characteristic of broken glass, and is the most common of all the non-metallic lusters.

Resinous and waxy luster

Resinous. . . . .luster is similar to the greasy appearance of resins and amber, and may grade into greasy luster. Waxy luster is what the name implies.

Pearly luster

Pearly luster is the luster of pearls, and is caused by innumerable thin, transparent plates piled one on another. Pile several sheets of clear cellophane on each other to secure the effect of pearly luster. This luster may be caused by innumerable cleavage cracks giving the same effect as a series of thin plates piled on each other.

Silky luster

Silky luster is the luster of silk, and is the result of a fibrous structure. It does not have a common occurrence but is very diagnostic when found.

Dull or earthy luster

Dull or earthy luster is similar to that of earth. It has a dull lifeless appearance.

CLEAVAGE

Cleavage is a quite important property of minerals and is defined as the tendency of certain crystals or crystalline material to break in definite directions.

One of the better known examples of cleavage is that of mica, which is perfect in one direction, and can be cleaved into very thin, paper-like sheets. Others, like calcite, break easily in three directions to yield a blocky figure. Instead of breaking clear across, one plane may break for a distance, discontinue, and be continued along another plane parallel to the first. Galena is also a good example of a mineral which shows such cleavage.

It is difficult to determine cleavage when the crystals are surrounded by other crystals. Tiny parallel cracks may give clues as to the cleavage. One should turn the specimen slowly this way and that, watching carefully all the time for brief flashes that are reflections from cleavage surfaces or crystal faces. The grains that "flash" can be studied under a hand lens and the cleavage details determined. Cleavage surfaces may be confused with crystal faces. Cleavage surfaces result when a mineral is broken, and the cleavage faces are repeated. Crystal faces exist only on unbroken surfaces. Two common types into which some cleavable minerals fall are:

Good cleavage in only one direction, such as mica. These minerals tend to break into sheets, or flat surfaces like a table top, folia, or scales.

Good cleavage in two directions.   Such minerals usually are found in elongated forms, or prisms, but other forms are also important.   The two cleavage surfaces may meet at right angles, or at oblique angles, and are characteristic of certain minerals.

Some minerals show cleavage in three (calcite) and four (fluorite) or six (sphalerite) directions

FRACTURE

Fracture is a term used to describe the manner, other than cleavage, in which minerals break.   It is not as diagnostic as cleavage, but it is important.   Descriptive terms are:

Conchoidal (pronounced kon-koid'-al: derived from the Greek konche, a shell, and eidos. form.)   It has a shell-like shape, similar to that on large, broken chunks of glass, and is characteristic of extremely fine-grained and glassy rocks and lavas.   Quartz is an example.   Sub-conchoidal fracture is poorly developed conchoidal fracture.

Hackly: is rough and uneven.   There is some regularity to the break but the surface is rough.

FORM

Most minerals are crystalline, and when free to form with no outside interference, take some regular, geometric shape, such as a cube, a prism, or any one of a host of other shapes.   The smooth surfaces are called crystal faces.

Examine some table salt with a hand lens!   Many "grains" will be seen as tiny cubes, having six faces, each of which is usually a square.   It is true that some of these faces may be slightly rounded, or perhaps elongated, but these are natural imperfections.   For our purpose, it can be said that table salt (the mineral halite) crystallizes in the cubic system.   The cube is the crystal form of halite.

Use of crystal form as a test is limited to specimens which have well-developed crystals, - otherwise this test should be avoided.   If a mineral crystallizes under crowded conditions, its growth is hampered by the growth of other minerals and the crystal faces may not be developed, and if only partial development of crystal faces took place, it would be very difficult to identify this mineral by its crystal form alone.   A great deal may be inferred about a mineral, however, if it occurs in its natural crystal form.   The six-sided quartz crystal (hexagonal) and the pyrite cube or its modifications are examples.

The six crystal groups, or systems, include: isometric, tetragonal, hexagonal, orthorhombic, monoclinic, and triclinic.   Each of these systems has a simple basic prism form that is generally considerably modified by the development of other forms.   Care must be taken not to confuse crystal faces with cleavage and fracture planes.   Cleavage and fracture surfaces develop when a mineral is broken, and occasionally may be parallel to a crystal face.

SPECIFIC GRAVITY

Pure minerals have a definite weight per unit volume (such as 1 cubic inch, 1 cubic foot, etc.). The ratio of this weight to the weight of an equal volume of water is the specific gravity of the mineral. Thus a specific gravity of four means that the mineral is four times as heavy as an equal volume of water.

The determination is based on the fact that a body immersed in water displaces an amount of water equal to its own volume. If its volume is four cubic inches it will displace four cubic inches of water, and the weight of this water divided into the weight of the object will give the specific gravity of the object. Furthermore, it is well known that an object "weighs" less in water than in air. It can be proven that this loss in weight is equal to the weight of the displaced water.

If an object is weighed in air and then immersed in water and the weight of the object in air is divided by the loss of weight when immersed in water, the answer is the specific gravity of that object. (Loss of weight in water is figured by subtracting the weight in water from the weight in air). It is therefore essential to have a reasonably accurate balance or scale with which to make weight determinations. Such a device is described in a following paragraph and is illustrated in figure 13.

A small portion of the pure mineral is placed in the pan, and the deflection of the beam is read at the right and recorded as the first reading, "A". Then the water is held so that the mineral and the pan are completely submerged, the mineral is placed in the second pan in the water and the deflection read and recorded as the second reading, "B". (Be sure that the pan does not touch the side of the glass). The second reading "B" is subtracted from the first reading "A", and the result of this subtraction is divided into the first reading "A".

$$\frac{\text{weight in air}}{\text{loss of weight in water}} = \frac{\text{weight in air}}{(\text{weight in air})-(\text{weight in water})} = \frac{A}{A-B} = \text{specific gravity}$$

The weighing device is made of two old hacksaw blades and some scrap lumber. The hacksaw blades are riveted or brazed end to end to form a flexible arm (1). A wooden base (2) about 3" wide and 10" or 12" long is mortised to support an upright (3). A slanting groove is cut in the side of (3) so that the free end of the blade (1) with its pan (4) is horizontal. A second upright (5) is mortised into the base (2) so that a graduated scale may be put on it. The pan (4) is made of a circular piece of thin metal, preferably copper, but a circular, smooth tin can top will serve, or the top with its rim, of a small baking powder can. Four holes are punched or drilled at quarter-points in this pan, and the pan is suspended from the flexible arm (1) by fine wire, such as window screen wire. Another similar pan (6) is suspended from the first so that it is always submerged in a water container (7).

The weighing device must have a set of graduations marked on (5) marked to indicate values, such as ounces, drams, or grams, etc. The calibration, as it is called, is accomplished by placing a weight of known value in pan (4) and marking the position of the flexible arm (1) on the paper. The first action is to mark the position of flexible arm (1) when no weights are in the pan; then mark the position of (1) as the smallest, and successively heavier weights are added

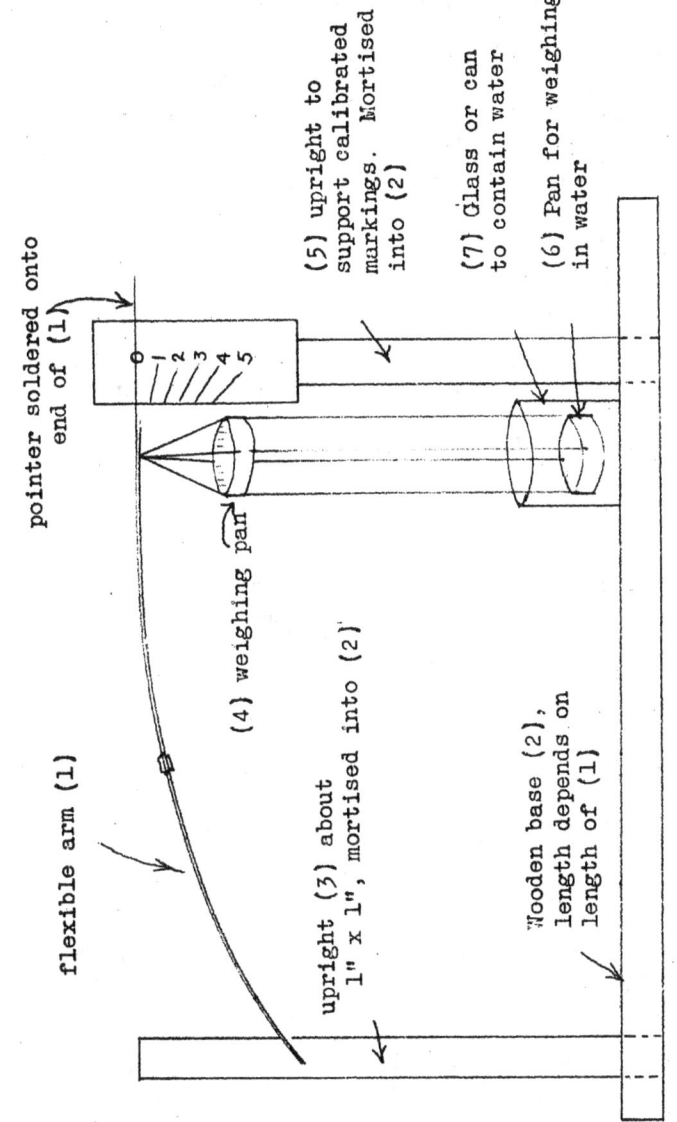

pointer soldered onto
end of (1)

flexible arm (1)

(4) weighing pan

upright (3) about
1" x 1", mortised into (2)

Wooden base (2),
length depends on
length of (1)

(5) upright to
support calibrated
markings. Mortised
into (2)

(7) Glass or can
to contain water

(6) Pan for weighing
in water

0 1 2 3 4 5

Figure 13. Weighing device.

to the pan.   Each mark on (5) should have the value of the weight placed after it.   When an object whose weight is unknown is placed in pan (4), the amount of deflection of (1) is read in ounces, or whatever unit was used for calibration.

The scale may be knocked down for packing by removing arm (1), pulling (3) and (5) from base (2), and stowing all parts to make a flat package.

HARDNESS

Hardness is one of the most diagnostic of all the tests.   It is based on the fact that materials differ in their degree of hardness.   Diamond is harder than glass, will therefore scratch glass, but glass will not scratch diamond.

A scale of hardness is called the Mohs scale, and ranges from 1 to 10.   The numbers indicate the comparative hardness; 5 is harder than 4, 7 is softer than 8.

Mohs scale number	Representative Material
1.	Talc, graphite.
2.	Gypsum
	Finger nail
3.	Calcite
	Brass pinpoint, copper penny
4.	Fluorite
5.	Apatite
	Knife blade, window glass
6.	Feldspar
7.	Quartz
8.	Topaz
9.	Corundum
10.	Diamond

Talc and graphite represent the softest minerals.   Gypsum is harder, but not necessarily twice as hard as talc; in other words, the numbers do not represent an arithmetical progression in hardness.   The finger nail has a relative hardness of about 2.5; it will scratch gypsum and talc, will not scratch calcite, but calcite will scratch all three.   An average knife blade is usually softer than window glass, but both will scratch apatite; feldspar will scratch both of them.

In actual practice, a knife blade is generally used and the ease or difficulty in making a scratch gives an idea of the comparative hardness.   If the groove is deep, the mineral is soft and may be below 3.   If the material is just barely scratched it may be 5.   If not scratched, the hardness is 6 or greater.   Let us assume that the mineral is scratched but not easily.   A brass pin will indicate if it is above or below 3.5.   Other testing agents may be used until the specific hardness is determined.

A small hardness set, consisting of small bits of mineral of specified hardness set in metal holders, is a very convenient testing device. Such a set may be purchased from Ward's Natural Science Establishment, Rochester, New York, for about $6.00. A fair general idea of hardness may be obtained by using the finger nail, a brass pin, a knife, and a piece of quartz.

Oftentimes a mineral surface appears to have been scratched but in reality the scratching tool has left a mark. If this is the case the streak will disappear when rubbed. Usually a scratch can be detected by dragging the finger nail across the surface. An examination of the surface with a hand lens before and after scratching will generally help to avoid errors, especially if there are already scratches on the surface. If the mineral appears to be scratched, look carefully to determine if it is a true scratch or whether the mineral crumbles so easily that you are actually grinding it to bits instead of scratching it. An example is sandstone: the grains are usually quartz with a hardness of 7 and therefore are not scratched by a knife blade, yet apparently they are scratched with ease. Actually, the cementing material may be quite soft and the sand grains are separated rather than scratched. Another case is that of galena in which the cleavage is so marked that a knife breaks out the cleavages instead of cutting them.

The minerals to be tested should be fresh and unaltered. Feldspar is 6 in the Mohs scale; it alters to kaolinite which is 2, and any gradation between may be found.

STREAK

Streak is the color of the powder of a mineral obtained by scratching the surface of the mineral with a knife or file, or, if not too hard, by rubbing it on an unglazed porcelain surface such as a streak plate. The resultant "streak" is finely powdered mineral.

Frequently streak color is different from the natural color of the specimen as viewed in daylight. An iron mineral, called hematite, has many forms and colors. Some are black, some are deep red, some are brick red; but in all cases the streak, or powdered mineral, is cherry to reddish brown. Calcite may have any color, yet its streak is always white. Graphite and molybdenite are so similar as to be easily confused; both are greasy black, nearly the same hardness, and have a cleavage like mica. Yet the streak of graphite is iron black while that of molybdenite is bluish to greenish black.

The most satisfactory way to obtain a streak test is to use a streak plate, as previously described. The plate may be used over and over again by wiping with a damp cloth to remove streaks.

In the event that it is impractical to secure a streak plate or piece of unglazed porcelain, the mineral may be finely powdered by crushing or by scraping with a knife. The resultant powder should be examined with a hand lens using a piece of white paper for a background.

ASSOCIATION OR OCCURRENCE

Certain minerals have definite rock associations, although rare exceptions to these rules may occur. For example, platinum and chromite are characteristically associ t d with ultra-basic rocks, that is, rock without free silica (quartz), and little feldspar. Chromite deposits occur so generally with serpentine that heir association is considered axiomatic. Many of these characteristic associati ns will be given under specific mineral descriptions.

ALTERATION

Many minerals have very distinctive alteration products, and these will be discussed under each mineral. Areas that have been subjected to extensive weathering are difficult to prospect, partly because the outcrops are soil covered, and pa tly because the minerals are so altered as to be unrecognizable unless a particular study has been made of the alteration products.

OTHER PHYSICAL PROPERTIES

There are several other physical properties which from time to time are of value in the identification of minerals. Some minerals (magnetite is the best example) are attracted by or even form a magnet. Others (halite or rock salt) have a definite taste. Certain minerals have a characteristic "feel" (talc, graphite). Others have a characteristic odor (kaolin has an "earthy" odor; barite a "fetid" odor when struck).

DESCRIPTION OF MINERALS

INTRODUCTION

Data used in describing the various tests for minerals have been taken from texts on mineralogy and from the experiences of the Department's staff.   Frequently it has been necessary to search several texts to get complete information for each mineral, and in some cases, to supplement it with personal experience. Differences of opinion about any particular test, such as specific gravity and hardness, have been settled by use of Dana's "Textbook of Mineralogy".

Phonetic spelling is used to indicate accepted pronunciation of mineral names. A certain amount of freedom with syllabification has been used to form syllable-words that have common usage and pronunciation.

PROCEDURE

The prospector or collector should devise some systematic method of identifying a specimen.   It is recommended that the mineral be examined and all of its characteristics determined before looking in a table or text.   A check list may prove valuable; it can be outlined in your notebook and the various items filled in as the determinations are made:

1. Color
2. Luster
3. Cleavage
4. Fracture
5. Form
6. Specific Gravity
7. Hardness
8. Streak
9. Occurrence

Then, start with some certain test say "hardness".   If the mineral has a hardness of "7", then all others are eliminated.   Next try "cleavage", and those not on the hardness-of-7 list are automatically eliminated.   Proceed with this method until the answer is found.

Should the prospector or collector perform one test and then try to determine the mineral without additional tests, there is danger that something important may be overlooked.   It is so easy to "want" a specimen to be a certain mineral, that it is difficult to make an impartial decision.

Finally, when you feel that the mineral is determined turn to the full description of that particular mineral and see if all points check.

DESCRIPTION OF MINERALS

AMPHIBOLE
(am'-fi-bole)

Amphibole is a name for a group of rock-forming minerals. Megascopically (seen with the naked eye or hand lens as opposed to seen through a microscope) they may be described and identified as a unit. Included in this group are:

Tremolite	Hornblende
Actinolite	Glaucophane
Nephrite (jade)	Crocidolite
Anthophyllite	Arfvedsonite

The amphiboles are usually dark-colored minerals that may occur in light-colored granitics. Usually they appear as slender crystals or needles, often singly, although they may occur in clumps, and look much like jet. The lighter colored varieties (tremolite, actinolite, crocidolite, etc.) frequently occur in bladed, fan-shaped masses, and alter to asbestos.

Amphiboles are complex silicates, and may contain calcium, magnesium, iron, sodium, and sometimes aluminum and fluorine. . . . . These are in such chemical combination that it is not practical to separate any of the elements for commercial production.

Color: varies with the amount of iron present, the more iron the darker the mineral. Tremolite may be almost white, actinolite is green, and hornblende may be dark green, brown, or usually black.

Luster: is bright and vitreous to somewhat pearly on cleavage faces. The acicular varieties may be silky.

Cleavage: is one of the better megascopic tests. Cleavage is good in 2 directions, parallel to the length of the crystal, and the two planes meet at angles of 125° and 55°, approximately those of a hexagon. (The hexagon angles are 60°). Note whether the cleavages develop surface angles that are separated by approximately two-thirds of a right angle, and if the surfaces have striations parallel to the length of the crystal.

Fracture: is uneven, and produces jagged sharp blades when broken across the cleavage.

Form: crystals are usually long and slender, have cross-sections that are nearly hexagonal (6-sided, 60° between faces) and the sides have striations (grooves) parallel to the length. The crystals in basaltic rocks may be shorter, but are still hexagonal in shape. Some varieties, notably tremolite and actinolite, are bladed or acicular (long needle shapes in slightly fan-shaped masses), and in radiated-columnar masses.

Specific Gravity: varies with the amount of iron, from 3 to 3.5.

Hardness: varies from 5-6; some specimens can be scratched with a knife and others cannot, but the general range is that of a knife-blade.

Streak: white to pale colors

Distinguishing Features: Amphibole may be confused with pyroxene, tourmaline, and epidote. It is distinguished from pyroxene mainly by cleavage, which is about 55° for amphibole and 90° for pyroxene, and by the brighter luster of the amphibole cleavages. Amphibole is frequently striated and pyroxene is not. Amphiboles are usually slender, pyroxenes are short and chunky. In many cases it is impossible to distinguish between them and then the name pyribole, a combination of both words, is used. Amphibole and tourmaline have similar cross-sections but tourmaline is usually triangular. Epidote has perfect cleavage in one direction, only. (See chart under pyroxene).

Occurrence: The amphiboles are found in igneous and metamorphic rocks. A general rule is to expect amphibole in the salic rocks like granite, and another similar mineral called pyroxene in the femic rocks like basalt. Examine specimens carefully, however, as amphibole (basaltic hornblende) may occur in basalts. Amphibole is common in gneisses and schists, and tremolite in limestone. Alteration amphiboles may alter to serpentine, chlorite, asbestos, and sometimes epidote and with continued weathering to limonite, carbonates, and quartz. Thus on greatly-weathered rock surfaces, only rusty-looking spots and holes may be left to show the former presence of amphiboles.

Similarities: Pryoxene. Tourmaline. Epidote.

Uses: The altered varieties, particularly asbestos, have commercial uses. (See asbestos).

ANGLESITE
(ang'-gl.i-site)

Anglesite, the lead sulphate ($PbSO_4$), is one of the minor ores of lead, containing when pure 68.3% lead. It is usually a white to grey mineral, found in the "oxidized zone" of lead deposits and therefore should have no appreciable depth. It may be a good indication of lead sulfide (galena) underneath. Anglesite may be found on timbers and walls of old lead mines where the sulfide has been taken into solution and deposited as a secondary mineral.

Color: white, tinged, gray, yellow, green, and sometimes blue.

Luster: adamantine to glassy; massive forms are dull; and transparent to translucent.

Cleavage: imperfect and not important.

Fracture: massive to conchoidal; brittle.

Form: crystals, banded masses.

Specific Gravity: 6.3.

Hardness: 3.

Streak white.

Distinguishing features: it is found in oxidized zones and is distinguished from
     cerussite (lead carbonate).   It does not effervesce in acid.

Occurrence: anglesite is a secondary mineral (found in the oxidized zone of lead
     deposits).

Similarities: cerussite; barite.

Uses: minor ore of lead.

ANHYDRITE
(an-high'-drite)

     Anhydrite and gypsum are closely related.   Both are essentially calcium sul-
fate ($CaSO_4$), but gypsum has in addition two molecules of water that anhydrite does
not have.   Anhydrite has no commercial use at present.

Color: white, grey, and tinted bluish and reddish.

Luster: cleavage faces have pearly to glassy luster, but massive varieties are
     dull.

Cleavage: 3 directions at right angles, producing cube-like forms.

Fracture: uneven, in massive forms; brittle.

Form: granular masses of sugar-like texture; crystals are rare. When crystals
     do occur they are orthorhombic.

Specific gravity: 2.9; somewhat higher than gypsum.

Hardness: $3-3\frac{1}{2}$, and higher than gypsum which is 2.

Streak: white.

Distinguishing Features: its higher specific gravity and hardness distinguish it
     from gypsum.   Its cube-like cleavage, when it can be distinguished, is also
     distinctive.

Occurrence: deposited from solutions; associated with gypsum; in sedimentary de-
     posits, such as cavities in limestones.   Large masses are found that can be
     quarried like limestone.   It also occurs as a metamorphic rock. Gypsum may
     change to anhydrite at depth.

Similarities: gypsum.

Uses: Anhydrite has little commercial use at present.   Experimental work
suggests that anhydrite may become valuable for certain uses when the oc-
casion arises.   See U. S. Bureau of Mines Information Circular 7049 for
details.

APATITE
(ap'-a-tite)

Apatite is a rock-forming mineral, sometimes difficult to identify megascopically.   It is calcium phosphate and is one of the principal sources of phosphate
in the anlyses of igneous rocks, but is not an important source of agricultural
phosphate in this country.

Color: Occurs as a light green, white, reddish-brown mineral with some variations.

Luster: vitreous to subresinous.

Cleavage: imperfect.

Fracture: uneven; brittle.

Form: hexagonal crystals and compact masses, sometimes granular.

Specific Gravity: 3.2.

Hardness: 5.

Streak: colorless.

Distinguishing Features: inferior hardness to beryl, about the same hardness as
glass, and subresinous luster.

Occurrence: in salic igneous rocks such as granite.   It is also a product of metamorphism, and further, it may be found in high temperature veins.

Similarities: beryl.

Uses: preparation of superphosphate for fertilizer.

ARGENTITE (silver glance)
(ar'-jen-tite)

Argentite is probably the most important ore of silver, and is chemically a silver sulfide (Ag_2S).   It is called "silver glance" by miners.

Color: Dark lead gray to dull black.

Luster: metallic.

Cleavage: unimportant, sometimes forming octahedral or cubic crystals.

Fracture: uneven to hackly.

Form: As massive crusts or coatings, and also in rough cubic crystals.

Specific Gravity: 7.3, high.

Hardness: 2-2$\frac{1}{2}$.   It is sectile, meaning that it can be cut easily, but is not malleable (flattened by a hammer blow).

Streak: Shining dark lead grey.

Distinguishing Features: Its low hardness, high sp. gr. and sectility are its most important distinguishing features.

Occurrence: As a vein mineral with other silver, lead, copper and iron sulfides.

Similarities: Chalcocite.

Uses: Important ore of silver.

ARSENOPYRITE (mispickel)
(are'-see-no-pie⌐rite)

Arsenopyrite is a silvery white to light steel mineral that may be confused with pyrite and other yellowish sulfides.   Chemical tests may be necessary to prove its identity, but a common field method often used is to rub the mineral vigorously and a sort of garlic odor is given off, indicating the presence of arsenic.

Chemical Composition: Iron- arsenic-sulfide ($FeAsS$)

Color: Silver-white to steel gray and often with a brassy tinge.

Luster: metallic.

Cleavage: not important.

Fracture: uneven.

Form: Orthorhombic crystals, often twinned, but usually granular compact masses
      and columnar forms.

Specific Gravity: 6.

Hardness: 5½-6.

Streak: Dark grayish-black.

Distinguishing Features: It is very similar to pyrite and may be distinguished
      if a garlic odor is noted when the mineral is rubbed or broken. It is only
      distinguished from smaltite by blowpipe tests.

Occurrence: Widespread; associated with the common sulfides.

Similarities: Pyrite; marcasite, smaltite.
Uses: Chief source of arsenic.

ASBESTOS
(as-bess'-toss)

Asbestos is a commercial term applied to certain minerals that have a well-
developed fibrous structure.   Chunks of these minerals may look much like serpentine but scratching or picking will separate the mass into individual fibers
that feel much like coarse "wool".

Asbestos is formed by alteration and is classed as a secondary mineral. Amphibole alters to asbestos (anthophyllite) under certain conditions and produces
a kind of fiber known commercially as non-spinning fiber.   This asbestos occurs
in radiating masses like a sheaf of broom-straws tightly held at one end.   The
fibers have unequal lengths, tend to be brittle, and cannot be woven, or "spun",
into cloth.   It has a lower market value than the "spinning" variety.

Alteration of serpentine may produce "cross-fiber" asbestos (chrysotile) of
"spinning" quality.   Commonly the asbestos occurs in veins, with roughly parallel
walls.   Fibers are of equal length, and as a rule are quite flexible and separate
into fine threads.   This material may be carded and spun much like regular organic fibers, and is used to prepare the asbestos cloth of commerce.   It rates the
highest price of all asbestos fiber.

References are:
      Bowles, Oliver: Asbestos, General Information: U.S.Bureau of Mines,
                      Information Circular 6817, January 1935.
      Bowles, Oliver: Asbestos: U.S.Bureau of Mines, Bulletin 403, 1937.

Color: variations of creamy gray, brown and green.

Luster: silky and glassy to dull.

Cleavage: perfect in some amphibole varieties and rare in others.

Fracture: splintery.

Form: acicular, slender crystals, in either radiating masses, or with fibers at
      right angles to wall rock.

Specific Gravity: 2.2 - 3.2.

Hardness: Varied, 2.5 - 5.5.

Streak: colorless.

Distinguishing Features: separates into fine slender fibers, that are somewhat flexible.

Occurrence: Asbestos is an alteration product of pyroxene or amphibole.

Similarities: Many zeolites have radiating structure, but the fibers are brittle.

Uses: various, mainly as fire-resistant products.

AZURITE (blue copper carbonate)

(az'-u-rite)

As its name suggests, this mineral has an azure-blue color. It contains a certain amount of chemically combined water and is properly called a hydrous copper carbonate ($Cu_3(OH)_2 \cdot (CO_3)_2$). It is characteristic of oxidized portions of copper deposits and ordinarily is superficial.

Color: azure-blue.

Luster: vitreous and velvety.

Cleavage: unimportant.

Fracture: conchoidal.

Form: azurite occurs in nodular groups of slender, needle-like acicular crystals and as crystalline coatings, frequently intergrown with green malachite.

Specific Gravity: 3.8.

Hardness: $3\frac{1}{2}$ - 4.

Streak: Light blue.

Distinguishing Features: The azure blue color, crystalline structure, and association with other copper minerals are all distinctive characteristics.

Occurrence: oxidized zone of copper deposits, associated with malachite.

Similarities: sodalite, lazulite, lapis lazuli (lazurite), hauynite.

Uses: minor ore of copper and as gem material.

BARITE (heavy spar)
(bare'-ite)

Barite has a non-metallic luster as well as a comparatively high specific gravity; it is used in the manufacture of white paints and barium salts. It may also be used to produce "green fire".

Chemical Composition: barium sulphate, $BaSO_4$.

Color: colorless, white, grey to tints of brown, blue, green, yellow, etc., depending upon the amount of impurities present.

Luster: vitreous.

Cleavage: perfect in 3 directions; two of the planes meet at right angles and one at an oblique angle.   (With calcite, all planes meet at oblique angles).

Fracture: uneven.

Form: tabular and platy crystals; globular and granular masses.

Specific Gravity: 4.5.   The mineral is sometimes called "heavy spar".

Hardness: 2.5 - 3.5.

Streak: white.

Distinguishing Features: Barite may be confused with calcite and dolomite. Barite has a higher specific gravity, does not effervesce in acids, and breaks into tabular forms.   Celestite and strontianite are best differentiated by flame tests.

Occurrence: found in separate veins and as a gangue mineral in deposits of lead, zinc, copper and iron sulfides, and also in limestone and sandstone beds.

Similarities: calcite, dolomite, celestite, strontianite, and scheelite.

Uses: in the manufacture of paint pigment; barium salts; "green fire"; and in the refining of sugar and as a heavy medium in well muds.

BAUXITE
(bokes'-ite)

Bauxite is an earthy mineral with a high alumina and low silica content. As a rule, chemical analyses are necessary to confirm this composition.   Characteristically, the mineral has a pisolitic or oolitic structure, with rounded, concretionary grains set in a clay-like groundmass.   However, bauxite occurs without such oolitic texture and may have a homogeneous, earthy appearance.   It may be either compact, or porous, rarely crystalline.

Bauxite is hydrous aluminum oxide with formula $Al_2O_3 \cdot 2H_2O$, of which 73.9 percent is alumina (Al_2O_3), and 26.1 percent is water (H_2O). All bauxite may contain such impurities as silica (SiO_2), iron sesquioxide (Fe_2O_3), titanium oxide (TiO_2), iron sulfide (FeS_2), iron carbonate ($FeCO_3$), and calcium carbonate ($CaCO_3$). Typical bauxite contains 2 to 10 percent silica, 10 to 30 percent water, 55 to 65 percent alumina, 0.5 to 25 percent iron sesquioxide, and 1 to 2 percent titanium dioxide. Diaspore ($Al_2O_3 \cdot H_2O$) has a similar composition to bauxite, but it is much harder and is usually brown or red in color.

Bauxite is the only commercial ore of aluminum in the United States, at present. It is the low silica content of bauxite rather than the high alumina content that is important, as silica is soluble in the same solutions as alumina. Many clays have high alumina content, even higher than commercial bauxite, but clays are not used in the United States for aluminum-production as the cost of extraction is greater. A German company has been producing aluminum from high-alumina clays for over a year, but no data are available as to comparative costs. [1]

Color: shades of brown, yellow, and red.

Luster: earthy.

Cleavage: none.

Fracture: earthy to conchoidal.

Form: occurs in clay-like masses which contain small "oolites", or "pisolites".

Specific Gravity: 2.5; fairly light of weight.

Hardness: soft; 1-3.

Streak: colorless

Distinguishing Features: clayey odor and clayey fracture. Chemical tests are usually necessary for exact identification.

Occurrence: an alteration product from the desilicification of clay.

Similarities: clay and laterite.

Uses: ore of aluminum, refractories and abrasives.

[1] Harder, E.C.: Industrial Minerals and Rocks, Seeley W. Mudd Series, A.I.M.E., 1937, pp.111-128.
Minerals Yearbook, U.S. Bureau of Mines, 1935, p.420.
Mineral Trade Notes, U.S. Bureau of Mines, vol.7 no.2 Feb.20 1938, p.2.

BENTONITE 1/   (Clays)
(ben'-ton-ite)

Bentonites are clays that usually have been derived from the weathering of volcanic ash and contain the clay mineral "montmorillonite" as their chief constituent.   Some varieties do and some do not increase in volume when placed in water; also plasticity varies considerably.

Color: light shades of gray, green and red.

Luster: earthy.

Cleavage: none.

Fracture: earthy and conchoidal like clay.

Form: massive microcrystalline aggregates of the clay mineral "montmorillonite".

Specific Gravity: variable.

Hardness: soft, 1-2.

Streak: colorless.

Distinguishing Features: dry surfaces adhere to the tongue; has a clayey odor when breathed on; and most varieties swell when placed in water.   Some varieties may "slack or break down" to fine particles in water.   A chemical analysis is necessary for an exact determination.

Occurrence: as a bedded deposit.

Uses: such clays have various uses, particularly those in which it is desired to take advantage of the swelling properties, namely: in dam construction where it is necessary to seal a pervious sand layer, and in well drilling to seal off a water stratum.   Other uses are as a filter, a de-colorizer, and as an ingredient of molding sands; as a bonding agent in combination with clays for making ceramic materials, as an adhesive and emulsifying agent, as a purifier of sewage, and as an ingredient in molding sand.   An important property, called thixotropy, has proven of value in the oil fields.   This is the property of a gel that permits it to solidify, and then become fluid, or viscous when agitated.   If such a material were pumped down an oil well hole, it would tend to seal off harmful areas, and when drilling is resumed, the gel under the drill would become viscous and flow out of the way. 2/

1/ Bechtner, Paul, Industrial Minerals and Rocks, Seeley W. Mudd Series, A.I.M.E.; 1937, chap.VI, pp.129-134.
    Davis, C.W.. and Vacher, H.C., Bentonite, Its Properties, Mining, Preparation, and Utilization: U.S.Bureau of Mines, Technical Paper 438, 1928.

2/ Mineral Trade Notes. U.S. Bureau of Mines,vol.8 no.5, May 20,1939, pp.28-29.

BERYL
(ber'-il)

The mineral beryl is a beryllium aluminum silicate ($Be_3Al_2(SiO_3)_6$) and is the source of the metal beryllium.   Theoretically, the mineral contains 14 percent beryllium oxide.   No commercial deposits of beryl have been found in Oregon, although occurrences have been reported but not verified.   Exceptional specimens may be used as gems; emeralds are a deep green variety; a lighter green is aquamarine: a beautiful rare pink beryl has been named morganite.

Color: commonly green, but sometimes white, yellow, pink. blue, or brown.

Luster: vitreous to greasy.

Cleavage: indistinct and imperfect.

Fracture: conchoidal or uneven; brittle.

Form: hexagonal crystals (six-sided prisms), or as compact columnar masses.

Specific Gravity: 2.7.

Hardness: $7\frac{1}{2}$ to 8, harder than quartz (H.7) and apatite (H.5) and softer than corundum (H.9).

Streak: none: it is too hard.

Distinguishing Features: beryl is harder than apatite and quartz, softer than corundum and chrysoberyl.   Otherwise, its hexagonal crystals are distinctive, when found.

Occurrence: in granite pegmatites, more rarely in mica schists and gneisses, and in calcite veins in limestone.   Large crystals are found in pegmatites.

Similarities: Quartz, apatite, corundum, and chrysoberyl.

Uses: source of beryllium metal, used in beryllium alloys.   Use of these alloys is discussed in Mineral Trade Notes, U.S.Bureau of Mines, Feb.20, 1939, vol.8 no.2, pp.3-5.   Pure forms when of good color are used for gems.

BIOTITE
(see Mica)

BLEACHING CLAYS

Bleaching clays are varieties similar to bentonite.   Whereas bentonite may have the property of absorbing large quantities of water, bleaching clays as a rule do not. They have a general waxy appearance, slake or disintegrate rapidly in water, generally lack plasticity, and have varying colors. These characteristics are not definite, however.   Since it is necessary to meet specifications of buyers, the reader is referred to the literature for these specifications:

> Bell, J. W., and Funsten, S. R., Industrial Minerals and Rocks, Seeley W. Mudd Series, A.I.M.E., 1937, chap.VII, pp.135-148 (contains excellent bibliography).

BORAX
(bo'-raks)

Borax is a sodium tetraborate, $Na_2B_4O_7.10H_2O$, which usually results from the repeated evaporation of intermittent shallow lakes.   Borax is one of a large number of borate minerals which serve as source material for the preparation of refined compounds of boron.   Other commercial borates are kernite, colemanite, and ulexite.

Color: Colorless and white when pure, commonly a dirty grey and blue.

Luster: Glassy and vitreous to dull.

Cleavage: Imperfect.

Fracture: Uneven

Form: Found as individual crystals, as aggregates of poorly-developed crystals, and as compact glassy masses.

Specific Gravity: 1.7.

Hardness: 2-2$\frac{1}{2}$.

Streak: White.

Distinguishing Features: Soluble in water and has a feeble sweetish alkaline taste.

Occurrence: In waters of saline lakes and in salt deposits resulting from the evaporation of such lakes.   Associated with other borates, nitrates and similar salts.

Similarities: Other borates, epsomite, melanterite, and alum.

Uses: Source of boron.   It is also finding use as a washing compound, as a water softener, and as a fireproofing agent for textiles.

BORNITE (Peacock ore)
(born'-ite)

Bornite is a copper iron sulfide, sometimes called "peacock ore" by miners, from its peculiar purple tarnish.   The color is similar to that which develops on the weathered surfaces of chalcopyrite.   Bornite occurs both as a primary and a secondary mineral.

Chemically, it is similar to chalcopyrite, being a compound of copper, iron, and sulfur (Cu_5FeS_4).   It contains 63.3 percent Cu, as compared with 34.5 percent in chalcopyrite.

Color: is a copper red and brownish bronze on the fresh surface, but readily tarnishes purple.

Luster: metallic.

Cleavage: none.

Fracture: usually uneven, but the purer more homogeneous varieties show conchoidal fracture; brittle.

Form: usually is found in massive form and in scattered (disseminated) specks through other copper ores; very rarely in rough cubic crystals.

Specific gravity: 5.

Hardness: 3.

Streak: grayish black.

Distinguishing Features: has a distinct purple tarnish.

Occurrence: in veins, in contact metamorphic deposits, and, less commonly, in disseminated deposits usually associated with pyrite, chalcopyrite, chalcocite, covellite and pyrrhotite.

Similarities: chalcopyrite, chalcocite, and covellite.

Uses: An important ore of copper.

CALAMINE
(kal-a-mine)
or
HEMIMORPHITE
(hem-i-more'-fite)

Calamine, or hemimorphite, is a silicate of zinc with water, having the chemical composition of $Zn_2(OH)_2SiO_3$.   It is very similar to smithsonite, the zinc carbonate, with which calamine may occur.

Color: colorless, white, and pale colors.

Luster: vitreous to pearly.

Cleavage: perfect in one direction.

Fracture: uneven to sub-conchoidal; brittle.

Form: massive, colloform, rounded forms, like smithsonite. Crystals are tabular, (flat, like a table, laminated).

Specific Gravity: 3.4 - 3.5.

Hardness: $4\frac{1}{2}$ - 5.

Streak: white.

Distinguishing Features: distinguished from smithsonite by cleavage parallel to the length of the crystals.

Occurrence: in the oxidized zone, usually derived from sphalerite, and associated with smithsonite.

Similarities: smithsonite.

Uses: ore of zinc.

CALCITE
(kal'-site)

Calcite, calcium carbonate ($CaCO_3$), is a common mineral. It may form as an alteration product in igneous rocks; it may be deposited from cold ground water or from hot spring waters; it may be deposited in sedimentary beds on the sea floor or it may form from the accumulation of calcareous sea shells which were relics of former marine life.

Some varied rock occurrences of calcite are:

Limestone, massive bedded dense and crystalline calcite.
Chalk, is a fine-grained, soft, porous limestone. Usually
    it is composed of the microscopic shells of marine organisms.
Marble, crystalline limestone resulting from metamorphism.
Coquina, is a limestone bed consisting of shells.
Travertine, is a cold ground water and hot spring deposit.
Oolite, is a limestone composed of small concretions ranging
    from the size of buckshot to the size of a pea. Sometimes
    it is reworked to form oolitic sandstone.
Iceland Spar, is clear, transparent calcite, free from chemical
    impurities and physical imperfections.

Color: colorless and transparent crystals and rhombs to dull opaque masses. It may be found in almost any color.

Luster: vitreous.

Cleavage: perfect in 3 directions. Cleavage directions are not at right angles, but form rhombohedrons that appear like a rectangular frame that has been pushed askew at one corner. The cleavage is so perfect that the large rhombs break easily into smaller individual rhombs.

Fracture : Indistinct or absent; brittle.

Form: Found in a great number of varied crystal habits; commonly as twins form-
     ing arrowheads and dog-tooth crystals.   Found as crystal coatings or druses
     in round cavities and veins and as imitative shapes such as stalactites and
     stalagmites.

Specific Gravity: 2.7.

Hardness: 3.

Streak: white.

Distinguishing Features: Calcite effervesces (bubbles) in cold dilute muriatic
     acid (HCl). Its cleavage is characteristic.

Occurrence: Widespread secondary and primary mineral, found in sedimentary beds
     and with igneous rocks.   Common as cavity and vein fillings deposited from
     circulating ground water.

Similarities: Aragonite, dolomite, siderite,anhydrite,rhodochrosite, magnesite,
     and gypsum.

Uses: Iceland Spar has optical use.   Calcite is the main constituent of lime-
     stone rock, which is used in agriculture; in the paper, cement, and chemi-
     cal industries.   It is used as road metal; as building stone; as monument-
     al stone; and as flux for the smelting and refining of metals.

CASSITERITE (Stream Tin),(Tin Stone)
(ka-sit'-er-ite)

Cassiterite is tin dioxide (SnO_2), commonly called "stream tin" or "tin-
stone", from its usual occurrence in placers.   Very little cassiterite is known
to exist in commercial quantities in the United States.   It is found in the New
England States; in the southern Appalachian Mountains; in South Dakota; in Riverside county, California; and in the placer gravels of Pine Creek, south of Baker,
Oregon.

Tin consumed in the United States is imported.

Color: brown or black and sometimes grey, red, or yellow.

Luster: submetallic to adamantine; crystals usually splendent.

Cleavage: imperfect.

Fracture: subconchoidal to uneven; brittle.

Form: it is found as crystals, as stream-worn pebbles, as mammillary nodules and
     sometimes as slender needles.

Specific Gravity: 7.

Hardness: 6-7.

Streak: white to gray and sometimes brownish from surface stains.

Distinguishing Features: high specific gravity, hardness, streak, and infusibility.

Occurrence: The most common primary origin is in quartz veins and granitic pegmatites.   About 2/3 of the world's production of tin is derived from placer deposits.

Similarities: Rutile, wolframite, zircon . . . . .and some varieties of garnet.

Uses: Source of tin.

CERARGYRITE (Horn Silver)
(sir-rar'-jer-rite)

Cerargyrite is a silver chloride (Ag Cl) and is known to the western miners as "horn silver" because of its softness and waxy or resinous appearance.   When exposed to light it darkens or tarnishes to a peculiar violet-brown color.

Color: white to pearl grey, green, and sometimes a violet tinge.

Luster: submetallic, waxy and resinous to adamantine.

Cleavage: none.

Fracture: unimportant, subconchoidal.

Form: massive, as thin crusts or coatings; rarely as cubic crystals.

Specific Gravity: 5.5.

Hardness: $1-1\frac{1}{2}$, highly sectile.

Streak: white, difficult to obtain.

Distinguishing Features: sectility, waxy appearance and sometimes the violet tinge.

Occurrence: It is a secondary mineral found in oxidized parts of silver deposits.

Similarities: Some copper carbonates.

Uses: Ore of silver.

CERUSSITE (White Lead Ore)
(ser'-oo-site)

Cerussite is the white lead ore, lead carbonate ($PbCO_3$) containing 77.5% lead. It is very brittle and commonly forms silky, milk-white masses of interlacing fibers, due to repeated crystal twinning.

Color: white, greyish brown, yellowish brown, and some darker tints.

Luster: adamantine and silky to dull.

Cleavage: in two directions, but not prominent.

Fracture: conchoidal; brittle.

Form: in compact masses, in fibrous and reticulated forms, and sometimes in crystals.

Specific Gravity: 6.5 (high).

Hardness: 3-3½.

Streak: uncolored or white.

Distinguishing Features: high specific gravity; bubbling in dilute nitric acid; and adamantine luster.

Occurrence: as a secondary mineral in the oxidized zones of lead deposits.

Similarities: anglesite, barite, witherite, and strontianite.

Uses: ore of lead.

CHALCEDONY
(kal-sed'-ony)
(varieties,-onyx, agate, ... plasma, etc.)

Chalcedony is a very fine-grained (cryptocrystalline) variety of silica or quartz, and usually represents deposition from low temperature solutions. Use of gem varieties has been the mainstay of the lapidary industry of the State of Oregon.   A variety of trade names is given to the different colored varieties of chalcedony and to those with peculiar impurities and structure differences. Some of the most common are petrified or agatized wood, banded agate (irregular color bands), moss agate (smoky, dendritic), sagenitic agate (acicular or needle-like inclusions), carnelian agate (reddish yellow), chrysoprase (green nickel oxide stain), enhydras (sealed water bubbles), onyx (straight color bands) and . . . . plasma (bright green or leek green; false jade).

Color: white with varied color tints; commonly a cloudy appearance.

Luster: waxy to dull.

Cleavage: none.

Fracture: conchoidal; brittle.

Form: as a secondary silica it may assume almost any form; usually massive and banded concretionary nodules.

Specific gravity: 2.6.

Hardness: 7.

Streak: none.

Distinguishing Features: hardness; conchoidal fracture. It shatters when broken and is finer grained than quartz and harder than opal.

Occurrence: a secondary mineral found as cavity and vein fillings in nearly all kinds of rocks, especially common filling amygdaloidal cavities in lavas; similar to zeolite occurrence.

Similarities: quartz, opal.

Uses: gem stone, and abrasive.

CHALCOCITE (copper glance)
(kal'-ko-site)

Chalcocite is often called "copper glance" by miners. Chemically it is cuprous sulfide, Cu_2S, and contains 79.8% copper when pure. It is the most important copper ore mineral of most of the world's largest copper mines.

Color: dark lead grey with a dull black or bluish tarnish which gives it the name of "sooty copper" or "copper glance".

Luster: dull metallic.

Cleavage: indistinct.

Fracture: hackly and granular to imperfect conchoidal.

Form: compact masses, sometimes granular, rarely in false-hexagonal crystals.

Specific Gravity: 5.7.

Hardness: $2\frac{1}{2}$-3, sub-sectile.

Streak: dark lead-grey to black.

Distinguishing Features: high specific gravity and association with other copper minerals; softer than bornite and tetrahedrite; less brittle than tetrahedrite; and more brittle than argentite.

Occurrence: commonly found in the enriched sulfide zone of copper deposits where it is clearly secondary in origin, but is known to occur as a primary mineral. It is frequently intermixed with bornite, covellite, chalcopyrite, and pyrite.

Similarities: tetrahedrite, argentite, and bornite.

Uses: high-grade ore of copper.

CHALCOPYRITE (fool's gold)
(kal'-ko-pie'-rite)

Chalcopyrite is a copper iron sulfide ($CuFeS_2$), containing when pure 34.5% copper. It has a yellowish metallic sheen and like pyrite has been mistaken for gold by inexperienced people. It is a common ore of copper in the United States and Canada.

Color: brass yellow to a silvery yellow; often tarnished with an irridescent purple color.

Luster: metallic.

Cleavage: none.

Fracture: uneven to sub-conchoidal; brittle.

Form: compact masses and crystals that look like tetrahedrons.

Specific Gravity: 4.2.

Hardness: $3\frac{1}{2}$-4.

Streak: greenish black.

Distinguishing features: it is softer than pyrite; is brittle while gold is malleable. Pyrrhotite is magnetic and chalcopyrite is not. Its surface tarnish has a purple sheen like bornite but its fresh surface is golden yellow.

Occurrence: with pyrite and metallic sulfides in many type of ore deposits and with various rock associations. Usually it is a primary copper mineral from which higher grade copper ores may be formed by alteration and replacement.

Similarities: pyrite, pyrrhotite, bornite, and gold tellurides.

Uses: most important ore of copper.

CHERT

Chert (or flint) is a term applied to hornstone and flinty rocks. It is a metamorphosed sedimentary deposit, occurring in stratified layers associated with limestone chalk and shale. Beds of chert 100 feet thick are found but usually such beds occur as oblong lenses and nodules distributed through sediments. The most common varieties are:

Flint, a black chert colored by organic matter, which disappears upon heating.
Jasper, a red and brown chert colored by iron oxides.
Novaculite (whetstone), a remarkably fine-grained, dense-appearing white chert which has been slightly metamorphosed and fractured.
Taconite, ferruginous chert of the Minnesota iron district.

Color: white, grey, green, red and black.

Luster: dull.

Cleavage: none.

Fracture: conchoidal (typical); brittle.

Form: rounded compact nodules and lenses to stratified beds; breaks with a sharp cutting edge.

Specific Gravity: 2.6.

Hardness: 7.

Streak: none.

Distinguishing Features: hardness, typical conchoidal fracture showing sharp cutting edges, and very fine-grained texture, often containing marine fossils or fragments.

Occurrence: in sedimentary beds; associated with limestone, dolomite, chalk, and shale. When pieces are rubbed or struck together, sparks fly and a peculiar odor is given off.

Similarities: chalcedony and quartz.

Uses: No typical modern use. It was used by ancient man to make arrow heads and flint tools.

CHLORITE
(klor'-ite)

The name chlorite includes an indefinite group of minerals that are always formed as alteration products from other minerals. Chlorites may resemble mica, particularly biotite, because of their green color. The general composition is a hydrous, magnesium, ferrous, aluminum silicate, with varying quantities of each.

Chlorites are found generally in foliated masses, that is, similar to a mat of leaves on a forest floor. The leaves are sometimes flat, but may be bent or curled.

Color: dark green.

Luster: vitreous to pearly.

Cleavage: one direction, basal, perfect but not as good as mica.

Fracture: splintery.

Form: foliated and scaly masses.

Specific Gravity: 2.6 - 2.96.

Hardness: 1-2$\frac{1}{2}$.

Streak: grayish green.

Distinguishing Features: scaly flakes not as elastic as mica, but more elastic than gypsum; distinct green color; . . . feel like that of talc.

Occurrence: an alteration product found in many igneous rocks; such minerals as biotite, amphibole, pyroxene, etc., alter to chlorite.

Uses: none.

CHROMITE

Chromite is the principal ore of chromium and is found in varying quantities in southwestern and northeastern Oregon. It has always been found associated with either greenstone, peridotite, or serpentine. The chemical composition is ferrous chromium oxide (FeO . Cr_2O_3) with varying amounts of magnesium and aluminum. To be commercial, chromite should contain better than 40% chromic oxide, or should be concentrated to that grade before shipping.

Color: black and brownish black.

Luster: metallic.

Cleavage: none.

Fracture: uneven.

Form: commonly in compact masses, sometimes granular, and as scattered black grains in the country rock.

Specific Gravity: 4.4.

Hardness: $5\frac{1}{2}$.

Streak: greyish brown and blackish brown.

Distinguishing Features: Sometimes it is feebly magnetic; has a dark brown streak, and is commonly associated with serpentine and magnetite.

Occurrence: found with the more basic igneous and metamorphic rocks, such as greenstone and peridotite or its altered equivalent, serpentine. Chromite deposits probably were formed as original magmatic segregations or were intruded into shear zones soon after solidification. [1]

Similarities: franklinite, magnetite, specular hematite, and ilmenite.

Uses: ore of chromium, which is used in the manufacture of stainless steel and other alloys; in making refractories, and in chemical industries.

CHRYSOCOLLA
(kris-o-kol'-a)

Chemically, chrysocolla is a hydrous copper silicate ($CuSiO_3.2H_2O$), containing when pure 36.1 percent copper and 20.5 percent water.

Color: pale blue to greenish blue.

Luster: vitreous, greasy to dull.

Cleavage: none.

Fracture: conchoidal to earthy, having an opal-like or enamel-like texture. Translucent varieties are brittle.

Form: found as . . . minutely crystalline mineral in veins and seam fillings, and sometimes as botryoidal nodules.

Specific Gravity: 2.2.

Hardness: 2-4.

Streak: white when pure; impure varities are light blue and light green.

Distinguishing Features: the light "glassy" blue color, inferior hardness, and the absence of any effervescence in dilute hydrochloric acid.

Occurrence: it is a secondary mineral often found in the leached and oxidized zones of copper deposits.

Similarities: turquoise, malachite, and genthite.
Uses: Minor ore of copper and gem stone.

[1] Allen, John Eliot, The Chromite Deposits of Oregon: State Department of Geology & Mineral Industries, Bull. no.8, 1938, pp. 29-31.

CINNABAR (Quicksilver)
(sin'-a-bar)

Cinnabar is mercuric sulfide (HgS).   It contains 86 2 percent mercury (quicksilver) when pure.   Prospecting frequently is done by means of panning either gravels or crushed samples since the mineral is heavy and like gold hangs back in the pan.

Color: commonly orange to crimson red and brownish-red; - a very distinct physical property.   Rarely, the color may be dark gray to black.

Luster: adamantine when pure; dull when in powdery masses and scattered through country rock.

Cleavage: rhombohedral (perfect in three directions), but this is not an important distinguishing property.

Fracture: Uneven.

Form: found as crystals disseminated through the rock and in massive and earthy forms.

Specific Gravity. 8; is exceptionally high and with color, the best distinguishing property.

Hardness: 2-2$\frac{1}{2}$.

Streak: scarlet or vermilion always a very brilliant red from the fairly pure mineral.

Distinguishing Features: its bright red color, scarlet streak, and high specific gravity.

Occurrence: cinnabar is usually a near-surface (low temperature hydrothermal) mineral, and may be deposited as irregular veins and seams, or as disseminated specks in rocks of various kinds and ages.   It is usually found associated with or near recent igneous intrusions.   Certain hot springs in volcanic regions are depositing cinnabar at the present time.

Similarities: hematite, cuprite, and realgar.

Uses: the most important source of quicksilver.

CLAY

The term "clay" is usually applied to certain earthy rocks whose most prominent property is that of plasticity when wet.   This permits molding into various shapes, which are retained when dry.

The clays consist of a group of secondary minerals intermixed with varying amounts of impurities (which consist mainly of quartz, calcite, limonite, muscovite, and rutile).   The clay minerals are (essentially) hydrous aluminum silicates, and may be distinguished from each other by optical methods.   The most common are:

Kaolinite	$Al_2O_5.2SiO_2.2H_2O$
Dickite	$Al_2O_3.2SiO_2.2H_2O$
Nacrite	$Al_2O_5.2SiO_2.2H_2O$
Halloysite	$Al_2O_3.2SiO_2.nH_2O$
Nontronite	$Fe_2O_3.3SiO_2.nH_2O$
Beidellite	$Al_2O_3.3SiO_2.nH_2O$
Montmorillonite	$(MgCa)O.Al_2O_3.5SiO_2.nH_2O$

Kaolinite is a common alteration product derived from the weathering of feldspars.   Dickite and nacrite are usually the result of hydrothermal action.   Both montmorillonite and beidellite are common as an alteration product of the glassy particles of volcanic ash and are thought to increase the bonding qualities of clay rock.

Clay particles are mostly below 0.002 mm. diameter, and are considered to be mostly of colloidal size, which permits them to stay in suspension in quiet water for a long period of time.

Clays may have an alumina content which is sufficiently high to arouse speculation as to their use as ores of aluminum.   As explained under Bauxite, it is the silica content, rather than the alumina content, that determines an ore of aluminum.   Clays usually have high silica content, and therefore are not considered as ores of aluminum in the United States.   A German company has been producing aluminum from high-alumina clays since 1938, but it is believed that the cost of production is greater than if bauxite were used.   Germany is deficient in bauxite reserves and must use   some aluminum-bearing mineral no' matter what the cost of extraction. [1]

Color: varied, from white, gray, pale yellow, blue and pink to black and dark red.   Most of the colored clays burn red, but, due to organic content, some black clays burn white.

Luster: dull and earthy.

Cleavage: none.

Fracture: uneven, the blocky clays break with a fairly even surface; some of the flint clays have conchoidal fracture.

Form: massive and bedded.

Specific Gravity: 2.6 .

Hardness: $1\frac{1}{2}$-$2\frac{1}{2}$, easily cut with the fingernail.

Streak: usually absent, but some of the red clays give yellow streak due to the limonite content.

[1] Mineral Trade Notes, U.S.Bureau of Mines, vol.7 no.2, Feb.20 1938, p.2.

<u>Distinguishing Features</u>: earthy appearance; clayey odor when breathed upon; sticks to the tongue when the specimen is dry; sticky, slippery and soapy feel when wet.

<u>Occurrence</u>: Clays may originate from various kinds of rocks, either by the ordinary process of surface weathering or by the action of solutions which may be either igneous or indirectly of surface origin. In both cases the alteration product is a residual clay. The removal of the clays so formed by various agents of erosion and their deposition elsewhere form a large group of transported clays, found in stratified layers and in pockets.

<u>Similarities</u>: bauxite, bentonite clay, siltstone, and shale.

<u>Uses</u>: as refractories and for making bricks, pottery, etc.; as a paper filler and a multitude of other uses. It is difficult to tell much about the value of a clay by visual inspection. Physical properties such as color when fired, fusibility, shrinkage, etc., are the most important. Oregon specimens may be submitted to one of the offices of the State Department of Geology and Mineral Industries for certain tests. 1/

1/ Mellor, J.W., Clay and pottery industries: vol.1, p.158, 1914. Lippincott.
    Ries, Heinrich, Clays, Their Occurrence, Properties, and Uses: John Wiley
        & Sons, 3rd ed., 1927.
    Ries, Heinrich, Clay, Industrial Minerals and Rocks: Seeley W. Mudd Series,
        A.I.M.E., 1937.
    Wilson, Hewitt, Ceramics; Clay technology: McGraw-Hill Book Co., 1927.
    Tyler, P.M., Marketing clay: U.S.Bureau of Mines, Information Circular 6998,
        1938.
    Wilson, Hewitt, and Treasher, Ray C., Refractory Clays of Western Oregon:
        Oregon State Dept. of Geology & Mineral Industries, Bull.no.6, 1938.

COAL

Coal is a term used to designate the more solid types of natural hydrocarbons; other forms of solid hydrocarbons are resins (amber), waxes (ozocerite); and asphaltum (gilsonite, wurtzilite, paraffin). Coal is in general the result of gradational change which has taken place in organic deposits, chiefly vegetable, by geologic processes, and form and composition of coal depend upon the extent of change caused by these processes. The resulting product consists of carbon, hydrogen, and oxygen with small amounts of nitrogen, sulfur and ash. Four general classifications representing different stages of change in the original deposit are recognized:

<u>Peat</u> is a brown to yellowish matted mass of interlaced, fibrous material, strongly resembling compressed tobacco. The remains of plant leaves, stems, roots, etc., in peat are still recognizable.

<u>Lignite</u> (brown coal) is a chocolate brown earthy material in which the texture and grain of wood and plant fibers are still distinct. It contains 55-75 percent carbon and burns readily with a smoky yellow flame and strong odor.

Bituminous coal is a compact, brittle coal breaking into cubical blocks with a gray-black to velvet-black color and brownish-black streak. It burns with a yellow flame, and very little of the original woody structure is evident. Carbon varies from 75 to 85 percent.

Anthracite coal is a compact, dense coal, iron-black to jet-black in color, with a vitreous luster, and conchoidal fracture. It usually has a hardness of more than 2, and burns with a pale blue flame. Carbon content varies from 80 to 95 percent.

Color: brown to black.

Luster: vitreous to submetallic and sometimes dull.

Cleavage: none.

Fracture: cubical or blocky with even surfaces and conchoidal fracture; earthy. Lignite, bituminous, and anthracite varieties are brittle.

Form: blocky jointed fragments with sharp edges and smooth fracture; often faintly banded.

Specific Gravity: 1.2 - 1.7.

Hardness: 1-2$\frac{1}{2}$.

Streak: brown, brownish-black, and jet-black.

Distinguishing Features: inflammability, light weight, brittleness.

Occurrence: in stratified layers; may be interbedded with sedimentary rocks, and may range from small seams up to beds several feet thick.

Similarities: other solid hydrocarbons; oil shale.

Uses: fuel; by-products. [1]

---

[1] Minerals Yearbook, U.S. Bureau of Mines, 1932-1936.
    Ladoo, R. B., The Natural Hydrocarbons: U.S. Bureau of Mines, R.I.2121 (1920).
    Libbey, F.W., Progress Report on Coos Bay Coal Field: Oregon State Department of Geology & Mineral Industries, Bulletin no.2, 1938.

COBALTITE (Cobalt Glance)
(ko-ball'-tite)

Cobaltite is a cobalt-arsenic-sulfide (CoAsS) that may contain as impurities small amounts of nickel and iron. The mineral has been found in the ores of several Oregon mines.

Color: silver-white with a reddish tinge; grayish black.

Luster: metallic.

Cleavage: perfect, cubic.

Fracture: uneven, brittle.

Form: granular massive, compact. Crystals shaped like those of pyrite, i.e.,
    cubes and pyritohedrons.

Specific Gravity: 6 -6.3.

Hardness: $5\frac{1}{2}$.

Streak: greyish black.

Distinguishing Features: cobaltite is found in crystalline form more often than
    is smaltite. Its cubic cleavage is distinctive, and it can be distinguish-
    ed from galena by its color and hardness.

Occurrence: in veins with silver and other cobalt and nickel minerals. In the
    the Standard Mine, Quartzburg district, Grant county, it occurs with chalco-
    pyrite, pyrrhotite, pyrite, and some smaltite.

Similarities: arsenopyrite; smaltite; pyrite; galena.

Uses: ore of cobalt.

COPPER MINERALS

There are many minerals which contain varying amounts of copper. Only a few
of those which are important as ores are described, under the following names:
Chalcocite, covellite, chalcopyrite, bornite, cuprite, malachite, azurite, tetrahedrite, and chrysocolla. Other copper minerals of less importance which are
not described here may be found in any standard textbook of mineralogy. Some of
these are: native copper, polybasite, enargite, atacamite, brochantite, linarite,
chalcanthite, turquoise and dioptase.

CORUNDUM (see p.98a)

COVELLITE.
(ko'-vel-ite)

Covellite is cupric sulfide (CuS) and may occur with chalcopyrite, bornite,
and chalcocite.

Color: indigo-blue or darker.

Luster: submetallic to resinous.

Cleavage: basal perfect, in one direction.

Fracture: none.

Form: massive; thin hexagonal plates.

Specific Gravity: 4.6.

Hardness: 1.5 - 2.

Streak: lead-gray to black.

Distinguishing Features: indigo-blue color; moistened with water turns purple.

Occurrence: secondary zone with other copper minerals.

Uses: ore of copper.

CUPRITE (Red Oxide, Ruby Copper)
(cue'prite)

Chemically cuprite is cuprous oxide (Cu_2O), containing when pure 88.8 percent copper. It occurs in the oxidized zone of copper deposits associated with other oxidized copper minerals, and often forms a coating on native copper.

Color: dark red and brownish red to ruby red.

Luster: adamantine, submetallic, or earthy.

Cleavage: imperfect.

Fracture: uneven; brittle.

Form: octahedral and cubic crystals, crystalline aggregates, and fine-grained masses.

Specific Gravity: 6.0.

Hardness: $3\frac{1}{2}$ to 4.

Streak: brick red and cochineal-red.

Distinguishing Features: association with other copper minerals; the absence of perfect cleavage; adamantine luster; streak is a brighter red than that of hematite.

Occurrence: In the oxidized zone of copper deposits, associated with native copper, malachite, azurite, limonite, etc.

Similarities: hematite, cinnabar, realgar.

Uses: minor ore of copper.

DIAMOND

Diamond is a chemically pure form of carbon; is one of the most desired of precious stones; is commercially by far the most important gem; is one of the few gems produced commercially by modern mining and milling methods. It cleaves perfectly parallel to the octahedron (8-sided, pyramid points). Diamonds have a very high refractive index, much higher than quartz and when rough quartz and diamond are dropped in a heavy oil medium, the diamond will show up much more plainly.than the quartz. It is transparent to X-rays, and its light ray dispersion is greater than that of any other mineral. [1]

[1] Ball, S. H., Industrial Minerals and Rocks, Seeley W. Mudd Series: A.I.M.E. 1937, pp.318-328.

Color: colorless and various shades of red, yellow, blue, green, and sometimes black. The black varieties are known as carbonado and bort.

Luster· adamantine to greasy.

Cleavage. perfect in four directions forming octahedrons.

Fracture: conchoidal.

Form: usually found as small loose crystals with rounded faces, curved edges, and a greasy to glassy appearance. Twin crystals are common.

Specific Gravity: 3.5.

Hardness· 10; hardest known substance.

Streak: none.

Distinguishing Features: extreme hardness; brilliant light ray dispersion, especially in the cut stones; greasy appearance, especially with rough stones. Diamonds have reportedly been found in Oregon stream gravels.

Occurrence: a most common occurrence is as a small grain or pebble in gravel, from which it is mined as is placer gold. The source rock of the placer deposits and the present source of approximately 32 percent of the production is from Kimberlite (blue ground), an altered ultra-basic rock.

Similarities: quartz, topaz, zircon, and corundum.

Uses: as a precious gem., Black diamonds (bort) are used as abrasives, in diamond drill bits, and various machine tools.

DIATOMITE (Diatomaceous Earth)
(dye-at'-tom-ite)

Diatomite is composed of the siliceous skeletons, tests, or frustules of diatoms, which are microscopic one-celled plants that live in both fresh and salt water. After dying they fall to the bottom of their habitat where the organic parts decompose and the silica skeletons may accumulate in beds of varying thickness. In time these beds become part of the sedimentary strata, usually mixed with small quantities of clay, carbonaceous matter, and other insoluble materials.

Chemically diatomite is essentially hydrous silica, containing 2 to 10 percent combined water and some impurities of organic and inorganic matter.

Diatomite is usually white when pure, frequently iron-stained along cracks, and very light of weight, because of the large amount of pore space. It is difficult to distinguish from fine-grained pumicite (volcanic ash) without the aid of a microscope. But when a small amount of pumicite or diatomite is ground between the teeth, the pumicite is distinctly gritty and brittle, whereas the diatomite is soft and crushes like chalk or clay.

Color: white and gray; colored by impurities.

Luster: dull to earthy.

Cleavage: none.

Fracture: roughly parallel to bedding planes and joint planes.

Form: massive bedded deposits which have a powdery chalk-like appearance.

Specific Gravity: 1.9 to 2.35 for pure hydrous silica. Diatomite is quite porous, however, and its apparent density is about 0.45 for blocks, and 0 12-0.25 for powder.

Hardness: 1 to $1\frac{1}{2}$, but the microscopic mineral particles are much harder, 4 to 6.

Streak: white.

Distinguishing Features: light weight, soft powdery feel between the fingers, and practically no grit or sand when crushed between the teeth.

Occurrence: it is found as stratified deposits in ancient lake basins, often interbedded with sedimentary rocks and volcanic lava flows.

Similarities: pumicite, clay, chalk, and weathered chert (tripoli). Chalk will effervesce in acid, while diatomite will not.

Uses: in general, its uses include heat and sound insulation; filtering medium; catalyst; absorbent; filler; mild abrasive; structural material; bleaching; chemical uses, etc. [1]

[1] Cummins, A. B., and Mulryan, H., Industrial Minerals and Rocks, Seeley W. Mudd Series: A.I.M.E , 1937, pp.243-260.
Moore, B.N., Non-metallic Mineral Resources of Eastern Oregon: U.S.G.S. Bull.875, 1937, pp.17-117.
Smith, W.D., Diatomaceous Earth in Oregon: Econ.Geol., vol.27 no.8, 1932, pp.704-715.
Hatmaker, P., Diatomite: U.S.B.M., Information Circular 6391. 1931.

DOLOMITE
(doe'-lom-ite)

Dolomite is a calcium magnesium carbonate, $CaMg(CO_3)_2$, with (theoretically) 54.4 percent calcium carbonate and 45.6 percent magnesium carbonate, or roughly half calcite and half magnesite (see discussion under calcite). The proportion of calcium to magnesium in dolomitic rocks may vary over a wide range. It is very abundant in certain regions, forming widespread thick sedimentary beds, such as those in the upper Mississippi and Missouri River valleys and in northeastern Washington.

Color: white or light gray when pure, often colored pink and bluish black by impurities.

Luster: pearly to vitreous.

Cleavage: forms rhombohedrons, similar to calcite. Cleavage is perfect in 3 directions at oblique angles, developing a figure which looks like a deformed rectangular prism. The sides, or faces of the dolomite rhombs are slightly curved; those of calcite are straight.

Fracture: subconchoidal; unimportant.

Form: massive beds or deposits and as crystal coatings or crystal aggregates in cavities, in veins, and as an accessory mineral.

Specific Gravity: 2.8 to 3.0, increasing with iron content.

Hardness: $3\frac{1}{2}$ to 4, slightly more than calcite (3).

Streak: none.

Distinguishing Features: only the powdered material effervesces in cold dilute HCl; the rhombic cleavage faces are curved; and sometimes there are twinning striations parallel to both diagonals; while in calcite the striations are parallel to the long diagonal. Rhodochrosite, barite, and siderite are heavier than the colored varieties of dolomite, and barite ($BaSO_4$) is insoluble in acids.

Similarities: calcite, magnesite, siderite, rhodochrosite, and barite.

Uses: refractory material for furnace lining; building and ornamental stone; filler for fertilizer.

EPIDOTE
(ep'-i-dote)

Epidote is a green mineral resulting from alteration or metamorphism of impure calcareous sedimentary rocks or igneous rocks containing much lime. The presence of this mineral is evidence of either contact metamorphism, hydrothermal alteration, or dynamic metamorphism.

Color: pale greenish yellow to pistachio green and greenish-black.

Luster: vitreous.

Cleavage: perfect in one direction.

Fracture: uneven.

Form: crystals usually elongated prisms, deeply striated; also fibrous, acicular, granular and massive.

Specific Gravity: 3.25 - 3.5.

Hardness: 6-7.

Streak: colorless.

Distinguishing Features: pistachio-green color and deeply striated prismatic crystals. See pyroxene for chart.

FELDSPAR GROUP
(feld'-spar)

Feldspar is a name applied to a group of rock-forming minerals which makes up about 60 percent of the igneous rocks. They are silicates with varying amounts of potassium, sodium, calcium, and aluminum, and are usually colorless or light colored. Varying amounts of impurities may impart some color. They have a white streak; cleavages in two directions, nearly at right angles; vitreous luster; hardness of 5.5 plus; and a specific gravity of 2.5 to 2.7.

Feldspars may be distinguished from quartz by their inferior hardness ($5\frac{1}{2}$-$6\frac{1}{2}$ as opposed to 7 for quartz), and cleavage. (Quartz has no cleavage). In a coarse-grained igneous rock, feldspars may be detected by the light which is reflected from their flat cleavage surfaces, and which gives such surfaces the appearance of a shellacked table top. When broken, ... feldspar surfaces . . . . . . develop a step-like appearance.

Feldspar crystals may have peculiarities of structure that result from what is known as twinning. They may look like crystal intergrowths, appear compound as if built of two halves, or, as in the plagioclase feldspars, exhibit multiple twinning which looks on cleavage surfaces like a series of parallel lines or fine striations.

Alteration: Slight alteration of the feldspars may cloud cleavage surfaces to a degree that they no longer are shiny and reflect light. In such a condition they might be overlooked and experience is necessary in order to distinguish those weathered minerals. Complete alteration results principally in the formation of kaolin which, if in deposits of commercial size, may be mined for china clay. A greenish tinge is due to the development of saussurite.

Alkali Feldspars

Orthoclase	$KAlSi_3O_8$
Microcline	$KAlSi_3O_8$
Anorthoclase	$(NaK)AlSi_3O_8$

Plagioclase (Soda-lime feldspars)

Albite $NaAlSi_3O_8$
   Oligoclase)
   Andesine   )   Mixture of albite
   Labradorite)   and anorthite
   Bytownite)
Anorthite $CaAl_2Si_2O_8$

Color: dominantly white and grey; sometimes colored (pink, green, yellowish
   green, cream yellow, and blue) by impurities. A play of colors (blue and
   green) is characteristic of certain labradorites.

Luster: vitreous and pearly, specially on the cleavage surfaces.

Cleavage: in 2 directions almost at right angles.

Fracture: uneven, poorly determined because of the good cleavage.

Form: common as single crystals, as twinned crystals, and as granular masses
   embedded in igneous rocks.

Specific Gravity: 2.5 to 2.7.

Hardness: $5\frac{1}{2}$ to $6\frac{1}{2}$.

Streak: white or grey, usually hard to obtain.

Distinguishing Features: crystalline structure, good cleavage, twin intergrowths.
   The plagioclase feldspars can be distinguished from the alkali feldspars by
   parallel twinning lines or striations.

Occurrence: they are found in all classes of rocks, but are most abundant in igneous rocks. The commercial potassium feldspars occur mostly as large crystals (single crystals may be larger than a man) in pegmatites. The plagioclase feldspars are more characteristic of the basic igneous rocks and basic
   lava flows.

Similarities: apatite, ... and rhodonite.

Uses: feldspars are very important in the ceramic industry; also used in the glass
   industry and for making scouring soaps. 1/

1/ Wilson, Hewitt, Some Feldspathic Materials of the Pacific Northwest: U.S.
   Bureau of Mines, Report of Investigations 2794, February 1927.
   Aitkens, I., Feldspar Gems (Amazon Stone, Moonstone, Sunstone): U.S. Bureau
   of Mines, Information Circular 6533, November 1931.
   Bowles, Oliver, and Lee. C.V., Feldspar: U.S. Bureau of Mines, Information
   Circular 6381, October 1930.
   Burgess, R.C.: Industrial Minerals and Rocks, Seeley W. Mudd Series, A.I.M.E.
   1937, pp. 261-282.

FLUORITE (Fluor spar)
(flu'-o-rite)

Fluorite is calcium fluoride having the chemical formula CaF_2.

Color: . . . . colorless, or some pale tint of violet, green, blue, etc.

Luster: vitreous; some specimens are fluorescent, that is, when viewed in ultra-violet light, the mineral emits a peculiar glow. Occasionally, the mineral is phosphorescent, that is, when treated with ultra-violet light, it will glow for a time after the light source has been removed.

Cleavage: its perfect octahedral cleavage in four directions is the most im-portant characteristic. . . . .

Fracture: sub-conchoidal; compact kinds splintery; brittle.

Form: One of the distinguishing features. The mineral crystallizes in the iso-metric (cubic) system and is usually found as cubes or sometimes as the octa-hedron, which is a modification of the cube. Also found as coarse or fine granular masses.

Specific Gravity: 3.0 to 3.2.

Hardness: 4.

Streak: white.

Distinguishing Features: harder than barite, celestite, and calcite, and softer than apatite; does not effervesce in cold HCl like calcite and strontianite. It forms cubic crystals, and has perfect octahedral (pyramidal) cleavage.

Occurrence: fluorite occurs as a gangue mineral with deposits of galena, sphal-erite, calcite, barite, and cassiterite; and as fissure veins in limestones and dolomite.

Similarities: calcite, barite, strontianite, celestite, and apatite.

Uses: manufacture of glass, enamels, and hydrofluoric acid. Pieces free from flaws may be used for optical lenses. As a flux in steel smelting, in the electrical extraction of aluminum; and in assaying.

GALENA (Lead Glance)
(ga-lee'-nah)

Galena, or lead sulphide (PbS), contains 86.6 percent lead when pure. It the most important ore of lead, and frequently contains enough silver to make it an ore of silver. Galena is a heavy greyish-black mineral, with perfect cubic cleavage having a silvery sheen on the cleavage surfaces.

Color: lead grey, often tarnishes to bluish white.

Luster: metallic.

Cleavage: very perfect in 3 directions at right angles (cubic). Smaller cubes may be easily broken from the larger cubes.

Fracture: undeveloped.

Form: commonly in cubes, less often as octahedrons modifying cubes, or as octahedrons. In ore bodies usually as cleavable masses.

Specific Gravity: 7.5 (heavy).

Hardness: $2\frac{1}{2}$.

Streak: lead grey.

Distinguishing Features: high specific gravity, cubic cleavage, and persistent lead grey color.

Occurrence: found as vein, contact metamorphic, and replacement deposits, associated with barite, fluorite, quartz, and the sulfides of copper, zinc and silver. Galena may alter to cerussite ($PbCO_3$), or to anglesite ($PbSO_4$).

Similarities: stibnite and argentite.

Uses: most important ore of lead and when silver-bearing it is an important source of silver.

GARNET
(gar'-net)

The name "garnet" is given to a group of minerals possessing similar physical properties and crystal forms, though their chemical compositions vary widely. The general chemical composition of the garnet group is $(CaMgFeMn)_3(AlFeCr)_2(SiO_4)_3$, and the more important members of the group are summarized as follows:

Grossularite $Ca_3Al_2(SiO_4)_3$, white, pale green, or yellow.
Pyrope $Mg_3Al_2(SiO_4)_3$, deep red or reddish black.
Almandite $Fe_3Al_2(SiO_4)_3$, common garnet, red or brownish red.
Spessartite $Mn_3Al_2(SiO_4)_3$, hyacinth red, sometimes a violet tinge.
Andradite $Ca_3Fe_2(SiO_4)_3$, brownish red, green, and black.

Color: white, green, brownish red, brownish yellow, red, and black.

Luster: vitreous to resinous.

Cleavage: usually absent, parting due to fracturing is common.

Fracture: uneven to sub-conchoidal; brittle.

Form: usually as distinct 12-sided crystals (dodecahedrons and trapezohedrons),
. . . . in the isometric (cubic) system of crystallization.
Sometimes as granular masses or as rounded or sub-rounded grains in sand.

Specific Gravity: 3.1 to 4.3.

Hardness: 6 to $7\frac{1}{2}$.

Streak: white; difficult to obtain.

Distinguishing Features: hardness, absence of cleavage, 12 sided crystal habit,
and high specific gravity, which is a little higher than similar appearing
hard gem minerals.

Occurrence: common as accessory minerals in a large variety of rocks all over
the world, but particularly common in contact deposits and in gneisses and
schists. Also found in crystalline limestone, serpentines, granite, and
occasionally in volcanic rocks; and because of its resistance to weathering
it is often found as rounded grains in river and sea sands.

Similarities: zircon, beryl, corundum, spinel, staurolite, epidote, and olivine.

Uses: most important in the manufacture of abrasives, especially abrasive-coated
papers and cloths. Flawless garnets may be used as gem stones or "jewel"
mounts.

GEMS (PRECIOUS AND SEMI-PRECIOUS)

There are a large number of precious and semi-precious stones and a complete
list is not given in this bulletin. Most of the commercial forms found in Oregon are varieties of silica, (quartz, chalcedony, opal) and are discussed in Bulletin no. 7, Gem Minerals of Oregon.

GOLD

Gold is a metal whose chemical symbol is Au, and it is found most commonly
in the free state, alloyed with a minor proportion of silver.

Color: deep to pale yellow, with sometimes a silvery tinge.

Luster: metallic.

Cleavage: none.

Fracture: hackly; malleable and ductile.

Form: usually finely disseminated through the containing rock, but megascopically
seen as rolled grains, flakes, wire stringers, nuggets, and rarely as isometric
crystals, rounded or subrounded as placer gold.

Specific Gravity: 15 to 19; pure Au is 19.3.

Hardness: 2½ to 3.

Streak: pale yellow, gold-yellow (same as the color).

Distinguishing Features: uniform pale yellow color, high specific gravity, mal-
        leability and ductility.   Pyrite and chalcopyrite are harder and are brittle.
        Golden colored mica is scaley and has a much lower specific gravity.

Occurrence: Gold is widely distributed in the earth's crust, occurring in quartz
        veins along with pyrite, chalcopyrite, galena, sphalerite, arsenopyrite; etc.;
        and in stream and beach placers along with such heavy minerals as magnetite,
        ilmenite, garnet, zircon, platinum, etc.

Similarities: pyrite, chalcopyrite, phlogopite (golden) mica.

Uses: basis of currency; in the arts.

GRAPHITE
(graf'-ite)

        Graphite is one of the softest minerals, and is a component of "lead" in lead
pencils.   When pure, it contains 100% of the element carbon, thereby having the
same chemical composition as the diamond; but often it has impurities of ferric
oxide, clay, etc.   The mineral has rather limited distribution in Oregon, mainly
in graphitic schists.

Color: steel grey to black.

Luster: metallic to dull.

Cleavage: basal; perfect in one direction, separating the mineral into thin
        scales, which are flexible and sectile.

Fracture: unimportant.

Form: usually found as foliated (like matted leaves) masses, as minute dissemin-
        ated scales, or as granular earthy masses.

Specific Gravity: 2.2.

Hardness: 1-2; soft greasy feel.

Streak: black or dark steel grey.

Distinguishing Features: its softness, greasy feel, black stain on the fingers,
        and its black streak, which is different from the greenish or bluish grey
        of molybdenite.

Occurrence: graphite is widespread in nearly all kinds of metamorphic rocks.   It
        has been found as beds, veins, and embedded masses, as laminae or scales in

granite, gneiss, crystalline schist, quartzite, limestone, and also in basic eruptive rocks.

Similarities: molybdenite, specularite, and solid hydrocarbons.

Uses. graphite is used in the manufacture of pencils, crucibles, lubricating oils, and stove polish.

GYPSUM (Selenite, Alabaster)

(gyp'-sum)

The mineral gypsum is calcium sulphate ($CaSO_4.2H_2O$), containing two molecules of water.   When one molecule of the water is driven off by heating, a powder - called Plaster of Paris - is formed.   If water is added to the powder, the mixture will harden when allowed to stand for a few hours, and form a cake or cement. Should the heating process be carried too far, all water is driven off and the mineral anhydrite ($CaSO_4$) is formed, which has little commercial use.

Gypsum occurs in various forms:

1.   Selenite: the clear, colorless crystalline form.
2.   Satin Spar: a fibrous variety.
3.   Alabaster: a fine-grained variety used for statuary.
4.   Rock gypsum, or gypsite: the more common commercial source.

Color: white, grey. pink, brown, etc.

Luster: pearly, silky, sometimes dull.

Cleavage: some varieties have perfect cleavage in one direction, imperfect in another, and are fibrous in a third.

Fracture: fibrous and sub-conchoidal; brittle.

Form: gypsum varies from foliated, fibrous, columnar, and granular masses to soft white powdery coatings (gypsite) and to single platy and fish-tailed crystals.

Specific Gravity: 2.3.

Hardness: $1\frac{1}{2}$ to 2.

Streak: white.

Distinguishing Features: softness (easily scratched with fingernail); one perfect cleavage in crystallized kinds; and non-effervescence in acids. To distinguish it from anhydrite the specimen should be carefully weighed, strongly heated for considerable period of time, and if it loses weight markedly, it is gypsum.

Occurrence: as bedded deposits associated with salt and limestone; formed directly by the evaporation of inland seas.   Secondary gypsum may originate through the interaction of sulphuric acid or other sulphate compounds with limestone or some other calcareous sediment.

Similarities: anhydrite, calcite, zeolites, talc, and mica.

Uses: for the manufacture of calcined gypsum (plaster of paris); for the manufacture of cement; for a soil stabilizer; for a paint filler; for a flux; and for ornaments.

HALITE (Rock Salt)
(hay'-lite)

Halite is the natural form of sodium chloride (NaCl), or common table salt, which is so essential to the health of man. It is also essential to the chemical industry, and fortunately the United States has extensive reserves.

Color: transparent or white when pure; often tinted yellow, pink, or blue.

Luster: vitreous.

Cleavage: cubic perfect and easily developed.

Fracture: conchoidal; brittle.

Form: halite is found as cubic crystals, as cubic fragments of larger crystals, and as granular to compact masses.

Specific Gravity: 2.1 to 2.6.

Hardness: $2\frac{1}{2}$.

Streak: white.

Distinguishing Features: its salty taste, perfect cubic cleavage, and softness.

Occurrence: most common in solid form as rock salt; found in beds or as salt domes (salt structures) associated with limestone, gypsum, anhydrite, as well as calcium and magnesium chlorides. Also occurs in sea water and in brines of inland or coastal lakes which have no outlet.

Similarities: such chlorides as sylvite (KCl) or carnalite ($KMgCl_3.6H_2O$), which are often found with halite. Fluorite is much harder and occurs differently.

Uses: for table salt and in the chemical, glass, and ceramic industries.

HEMATITE (Specular Iron, Red Iron Ore, Kidney Stone)
(hem'-a-tite)

Hematite is an iron (ferric) oxide (Fe_2O_3), containing when pure 70% iron. It is the most important ore of iron in the Great Lakes region and in the southern Appalachian Mountains. Some hematite occurs in Columbia county, northwest of Portland, Oregon, but most of the iron ore of this region is limonite. Hematite is found in many forms:

Specular iron (specularite) is steel grey, dark brown, or brownish black foliated or flaky hematite.

Kidney ore (red hematite) is often red or dark brown, occurring in columnar, fibrous or reniform (mammillary) masses.

Oolitic hematite is the red sedimentary iron ore of the southern Appalachian Mountains. It looks like cemented buckshot, and consists of small, rounded concretionary grains.

Massive hematite is important in the Lake Superior region.

Color: steel grey, brown, red, and black.

Luster: metallic, sometimes dull.

Cleavage: none.

Fracture: earthy, hackly, and sometimes conchoidal.

Form: as described above, the form is variable.

Specific Gravity: 5.0.

Hardness: $5\frac{1}{2}$ to $6\frac{1}{2}$.

Streak: a bright brownish red or dark red, and is a distinctive physical property.

Distinguishing Features: brownish red streak, non-magnetic character, and low water content or complete absence of water.

Occurrence: both as disseminations and massive with various rock associations as contact metamorphic deposits; as residual alteration deposits; as metasomatic replacements of cherty iron carbonate; and as hematite schists in metamorphic rocks. It is very widely distributed in small deposits.

Similarities: limonite, magnetite, chromite, ilmenite, and franklinite.

Uses: the most important iron ore.

ILMENITE
(ill'-men-ite)

Ilmenite is titaniferous iron oxide ($FeTiO_3$). It is very common in coarse-grained basic igneous rocks and has many characteristics similar to magnetite, hematite, and chromite.

Color: iron black.

Luster: sub-metallic and metallic.

Cleavage: none.

Fracture: conchoidal; brittle.

Form: it is found as tabular crystals and flat plates; as disseminated grains in compact masses; and as loose rounded sand grains.

Specific Gravity: 4.5 to 5.

Hardness: 5-6.

Streak: black to brownish black, but darker than hematite. Its composition is similar to hematite except that one molecule of titanium has replaced one molecule of iron.

Distinguishing Features: brownish black streak. Non-magnetic or weakly magnetic. Chromite occurs in octahedrons.

Occurrence: in basic coarse-grained igneous rocks and in black sands; associated with magnetite, hematite, chromite, pyrite, garnet, etc.

Similarities: magnetite, hematite, franklinite, chromite, wolframite, etc.

Uses: ore of titanium, used in paints, and in steel alloys.

LIMONITE (Brown Hematite, Bog Ore, Yellow Ocher)
(lie'-moe-nite)

Limonite is hydrated ferric oxide ($2Fe_2O_3.3H_2O$), containing when pure about 60 percent iron and about 14 percent water.

Color: ocher-yellow to yellowish brown, to brown.

Luster: in compact varieties often silky to sub-metallic, but generally dull or earthy.

Cleavage: none.

Fracture: uneven, conchoidal.

Form: variable, occurring as botryoidal, nodular, pisolitic, and porous.

Specific Gravity: averages 3.8.

Hardness: 1 to 5; about $5\frac{1}{2}$ when pure.

Streak: yellowish brown.

Distinguishing Features: its yellowish brown streak, its high water content, and its structure or form.

Occurrence: . . . . as a secondary mineral formed by the breaking down of iron-bearing minerals and rocks. It has the same general chemical composition as common rust which forms on metallic iron when exposed to a humid atmosphere. It may occur as a sedimentary bedded deposit, as a metasomatic replacement of limestone, and as pseudomorphs.

imilarities: hematite and other iron oxides, manganese oxides, bauxite.

Uses: as an ore of iron and as a yellow pigment.

MAGNESITE
(mag'- neh- site)

Magnesite is magnesium carbonate ($MgCO_3$), containing 47.6 percent magnesia
(MgO). Sometimes it is hard and tough owing to admixed silica. In sedimentary
deposits it may be mechanically mixed with calcite and dolomite, which have very
similar physical properties; sometimes it contains a little siderite. Extensive
deposits occur in Stevens county, Washington, and in California and Nevada.

Color: snow-white to grey; also reddish or brown.

Luster: vitreous to dull.

Cleavage: in the crystalline varieties, rhombic like calcite.

Fracture: conchoidal in fine-grained compact masses.

Form: most common in compact, porcelain-like masses; sometimes coarsely crystal-
       line and granular as in the Washington deposit.

Specific Gravity: 3.0.

Hardness: $3\frac{1}{2}$ to $4\frac{1}{2}$, sometimes harder when silica is admixed.

Streak: white.

Distinguishing Features· its fairly high specific gravity and hardness, its
       toughness, its very fine-grained texture; and its mild effervescence in cold
       dilute HCl.

Occurrence: the larger deposits are lenses in sediments formed by hydrothermal
       replacement of dolomite; some deposits are formed by the alteration of rocks
       rich in magnesia, and are associated with metamorphic minerals such as talc,
       chlorite, serpentine etc.

Similarities: calcite, dolomite, and rhodochrosite.

Uses: the greatest demand is for refractory purposes; other uses are as a chem-
       ical accelerator in rubber, and for cements.

MAGNETITE (Magnetic Iron Ore, Lodestone)
(mag'- neh-tite)

Chemically magnetite is iron oxide Fe_3O_4, sometimes written $FeO.Fe_2O_3$. It
is usually strongly magnetic and will deflect a compass needle.

Color: dark grey to iron-black.

Luster: metallic, sometimes rather dull.

Cleavage: not distinct, but it breaks or parts parallel to the octahedral faces.

Fracture: sub-conchoidal; brittle.

Form: most commonly occurs as small crystalline or non-crystalline grains either embedded in a massive rock or loose in black sands. Sometimes found as octahedral crystals and as granular masses.

Specific Gravity: 5.1

Hardness: $5\frac{1}{2}$ to $6\frac{1}{2}$.

Streak: black.

Distinguishing Features: magnetite is strongly magnetic, that is, it is attracted to a magnet, and the "lodestone" variety will act as a magnet. It has a black streak which differentiates it from chromite or hematite. Both franklinite and ilmenite have a similar streak, but are only weakly magnetic.

Occurrence: magnetite is very widely distributed, being very common in crystalline basic igneous rock as small grains and as segregated masses. It is also common in highly metamorphosed rocks and in contact metamorphic deposits. Magnetite is usually the principal constituent of "black sand".

Similarities: ilmenite, franklinite, hematite, and chromite.

Uses: an iron ore.

MALACHITE
(mal'-a-kite)

Malachite is a hydrous copper carbonate, $Cu_2(OH)_2CO_3$, containing when pure 57.4 percent copper and 8.2 percent of chemically combined water.

Color: emerald green.

Luster: vitreous and velvety, often dull or earthy.

Cleavage: unimportant.

Fracture: uneven, sometimes sub-conchoidal; brittle.

Form: commonly mammillary or kidney-shaped with banded structure; also massive; stalactitic; crystals rarely distinct, but when present their form is slender or acicular.

Specific Gravity: 3.9 - 4.0.

Hardness: $3\frac{1}{2}$ to 4.

Streak: pale green.

Distinguishing Features: uniform emerald green color; effervescence in cold dilute HCl, fibrous or acicular crystal structure.

Occurrence: malachite results from oxidation of other copper minerals and is found in the upper oxidized zone of copper deposits.

Similarities: weathered azurite, chrysocolla, turquoise, and smithsonite.

Uses: ore of copper; some varieties used as gem stones.

MANGANESE MINERALS
(man'-gan-ease)

There are six fairly common manganese minerals.   Each is described under a separate heading, as follows: manganite, psilomelane, pyrolusite, rhodochrosite, rhodonite, and wad.   The last is a soft mixture of several of the other oxides.

MANGANITE
(man'-gan-ite)

Manganite is a hydrous manganese oxide, $Mn_2O_3.H_2O$ containing when pure 62.4 percent manganese.   It is a secondary mineral commonly filling veins and cavities.

Color: iron-black or dark grey.

Luster: submetallic.

Cleavage: when present, in one or two directions parallel to the length of the crystals.

Fracture: uneven.

Form: in columnar and fibrous masses and in prismatic crystals, which are vertically striated and often grouped in bundles.

Specific Gravity: 4 3.

Hardness: 4.

Streak. reddish brown to black

Distinguishing Features: its crystalline structure harder than pyrolusite, softer than psilomelane; streak.

Occurrence: most commonly as a secondary mineral filling veins and cavities associated with barite and calcite and also found in residual clays with psilomelane.

Similarities: psilomelane, pyrolusite.

Uses: an ore of manganese.

MERCURY (Quicksilver)

Metallic mercury, or quicksilver, occasionally occurs free in nature as small globules, usually associated with cinnabar, in quicksilver deposits.

Mercury minerals are: mercury, cinnabar, metacinnabarite, and calomel.

MICA MINERALS

These minerals are included in a group which have similar physical properties with the general chemical formula R Al SiO_2H_2O, (the "R" standing for K, Mg, Fe, Li or Ti), and are as follows: muscovite, biotite, phlogopite, vermiculite, and lepidolite.   Their physical characteristics:

Color: various; water clear to dense black.

Luster· cleavage surfaces are pearly to vitreous.

Cleavage: perfect basal cleavage; mica may be split into thin sheets.

Fracture: seldom developed.

Form: when crystallized, most micas form squat pseudo-hexagonal prisms.

Specific Gravity: 2.7 - 3.1.

Hardness: 2.5 - 3.5.

Streak: white to pale green, usually lighter than the solid mineral.

Distinguishing Features: perfect basal cleavage.

Occurrence: mica is found in nearly all of the acid igneous rocks such as gran-
     ite, syenite, diorite, and pegmatite.   Commercial sheet or book mica is
     found in granitic pegmatite.   Mica flakes are a common constituent of many
     sands.   Mica is also generally contained in schists.

Similarities: chlorite, gypsum, talc.

Uses: muscovite and phlogopite are used as insulating materials in electrical
     apparatus.   Phlogopite can withstand up to $1000°$ C.   Vermiculite after
     heat treatment is used for insulating purposes in building construction.
     Powdered vermiculite and muscovite are used in roofing compounds.

The mica minerals as outlined above have individual characteristics as follows:

Muscovite (H_2K $Al_3(SiO_4)_3$), is usually practically colorless, but is sometimes
     grey, light green and yellow.   It is commonly found in foliated masses and
     scales.   In pegmatites it may occur as large books.   Muscovite is usually

flexible, and resistant to weathering. The fine-grained variety of muscovite is called sericite, and gives a golden sheen to the rock in which it appears.

Biotite: ($(HK)_2(MgFe)_2Al_2(SiO_4)_3$) is the most common form of mica, and is usually dark green or black in color. It has less flexibility and is less resistant to weathering than muscovite.

Vermiculite: this mineral, which resembles biotite, is considered to be a hydrated form of biotite. It possesses the peculiar property of exfoliating or expanding to a remarkable degree when heated, even with a match. This characteristic is due to the formation of steam between the separate layers.

Phlogopite (Magnesian mica): this is a brown or amber colored mica, composition of which is closer to that of biotite than to muscovite. Its color appears golden on cleavage surfaces and is sometimes mistaken for gold flakes when finely divided in sand.

Lepidolite (lithia-iron mica): this mineral is rather rare. It has been found in pegmatites and is mined as an ore of lithium. It is easily recognized by its rosy lavender color, which is almost unique among minerals.

MOLYBDENITE
(moe-lib'-de-nite)

Molybdenite is molybdenum sulfide (MoS_2); it has an appearance very similar to that of graphite.

Color: bluish lead grey.

Luster: metallic.

Cleavage: perfect in one direction, almost micaceous. The sheets are slightly flexible and sectile.

Fracture: none.

Form: hexagonal, but molybdenite is seldom found in crystal form. It usually occurs as foliated masses or as disseminated (scattered) scales.

Specific Gravity: 4.7 - 4.8.

Hardness: $1-1\frac{1}{2}$, slightly greater than graphite; easily scratched with the finger nail.

Streak: greenish to bluish black, but makes a bluish-grey to greenish grey mark on paper.

Distinguishing Features: molybdenite is very similar to graphite, but may be distinguished particularly by its greenish to bluish streak on a streak plate. Graphite streak is lead colored. Molybdenite also has higher specific gravity and a brighter color.

Occurrence: found in the more siliceous rocks, such as granites and pegmatites, in veins, in contact deposits, and as disseminations in country rock.

Similarities: graphite.

Uses: ore of molybdenum.

MONAZITE
(mon'-a-zite)

Monazite is . . . .a phosphate of the cerium metals with a formula which may be represented as $(Ce,La,Nd,Pr)PO_4$. but thoria and silica are nearly always present. The importance of monazite commercially is due to the thoria content.

Color: honey yellow, brown, reddish.
Luster: resinous, translucent in grains; large crystals in rocks dull to opaque.
Cleavage: parting (similar to cleavage) may be prominent in one direction.
Fracture: conchoidal to uneven.
Form: usually in the form of grains in various sands; sometimes in prismatic crystals and in masses in gneissoid rocks and pegmatites.
Specific Gravity: 4.7 - 5.3.
Hardness: 5.
Streak: grains usually too small to give a streak.
Distinguishing Features: color, mineral association. Monazite possesses radio-active properties strong enough to affect a photographic plate. If an un-exposed photographic plate or film is wrapped in black photographic paper, without exposing it to the light, and a key, coin or other small, distinctly shaped metal object having a small amount of the supposed monazite sand on top of it is placed on the outside of the black paper, the radioactive rays given off by the monazite will pass through the black paper but not through the metal object, and the photographic negative will be fogged except where covered by the metal object. The plate or negative is then developed in the usual manner; a fogged plate will indicate the presence of radioactive mat-erial.
Occurrence: in some sand deposits, and in gneisses and pegmatites.
Similarities: sphene (titanite) and zircon; usually optical tests are necessary to detect the presence of monazite in sands.
Uses: as a source of thorium which when made into a nitrate is used in incandes-cent mantles.

MUSCCVITE
(muss'-ko-vite)
(see Mica)
NICCOLITE (Copper Nickel)
(nick'-ko-lite)

Niccolite is nickelarsenide, having the chemical formula NiAs.   It usual-ly contains a little iron, cobalt, sulphur, and antimony.

Color: pale copper red.
Luster: metallic.
Cleavage: none.
Fracture: uneven.
Form: nearly always massive.
Specific Gravity: 7.33 - 7.67.
Hardness: $5 - 5\frac{1}{2}$.

Streak: pale brownish black.
Distinguishing Features: color, specific gravity, mineral association.
Occurrence: found associated with native silver and smaltite and more rarely
     with copper and other metallic sulphides.
Similarities: pyrrhotite.
Uses: ore of nickel.

NITRATES

Nitrates include such minerals as contain the nitrate radical, and are
usually soluble in water.   The principal mineral,nitratine, is commonly known
as soda niter.   Chemical composition: usually sodium nitrate, although impurities may also be present.

Color: white or colorless.
Luster: earthy.
Cleavage: practically identical with calcite, that is, at angles of $73°30'$.
     This is known as rhombohedral cleavage. (See Calcite).
Fracture: not apparent.
Form: nitratine crystallizes in rhombohedrons (see Calcite) very similar to cal-
     cite. Usually, however, the mineral is massive and the crystal form is dis-
     tinguished with difficulty. As the mineral is soluble in water, if the wa-
     ter is evaporated, crystals may form that will permit identification.
Specific Gravity: $1\frac{1}{2}$ - 2.
Hardness: $1\frac{1}{2}$ - 2.
Streak: white.
Distinguishing Features: its cool, saline taste.
Occurrence: nitratine is soluble in water. Therefore deposits are usually found
     in arid regions, such as southeastern Oregon. In that region, it is asso-
     ciated with rhyolites, and occurs in cavities as incrustations. There have
     been efforts made to mine the nitrates on a commercial scale, but they have
     met with no great success because the deposits are not extensive.
Similarities: none.
Uses: nitratine or nitrates are valuable for fertilizers and in the manufacture
     of potassium nitrate.

OLIVINE
(ol'-i-vene)

Olivine is a silicate with varying amounts of iron and magnesium and a general formula of $(Mg, Fe)_2SiO_4$. It is the chief rock-forming mineral of the ultra-basic rocks, making up a large percentage of such coarse-grained types as peridotite. Pure olivine rock is called "dunite". Olivine appears as crystals in
many basalts.

Color: yellowish green to bottle green.   Weathering may give it a reddish to
     rust yellowish green color.
Luster: vitreous.
Cleavage: not distinct.
Fracture: conchoidal, brittle.
Form: usually found in disseminated masses (grains scattered through the rock) or
     segregations of granular masses. The crystals are orthorhombic and tabular.
Specific Gravity: 3.3.
Hardness: $6\frac{1}{2}$ - 7.
Streak: colorless.
Distinguishing Features: its olive-green color and the absence of cleavage.
Occurrence: basic, or femic, igneous rocks such as peridotite and basalt. Segre-
     gations of sizeable masses may be commercially important.

Similarities: epidote.

Uses: large masses of dunite have important use in the manufacture of refractories. Low iron content is required, as presence of iron seems to cause its fusion at lower temperatures. Clear crystals are used as gems, under the name of peridot.

OPAL
(O'-pal)

Opal is an hydrous form of silica. When pure it may be used as a gem. More frequently it is mixed with clay and other impurities. It is silicon dioxide with water ($SiO_2.2H_2O_x$).

Color: any color. Precious opal has a play of colors due to minute fracturing.

Luster: greasy to pearly.

Cleavage: none.

Fracture: conchoidal. Very brittle, the most fragile of all gems.

Form: amorphous (non-crystalline) silica in lumps and masses.

Specific Gravity: 2 to 2.3.

Hardness: $5\frac{1}{2}$ to $6\frac{1}{2}$.

Streak: none.

Distinguishing Features: its softness and lower specific gravity distinguish it from quartz and chalcedony, as does its greasy luster.

Occurrence: opal is deposited from alkaline solutions and has no crystalline structure (amorphous). It may alter to chalcedony through loss of water and by becoming cryptocrystalline (microscopic crystals).

Similarities: chalcedony.

Uses: certain varieties are important gem materials.

ORPIMENT
(or'-pih-ment)

Orpiment is an arsenic sulfide, with a lemon yellow color. Some realgar is usually present. Its color and association with realgar make it an easy mineral to identify. (See Realgar).

Color: lemon-yellow.

Luster: dull.

Cleavage: not distinct.

Fracture: none.

Form: foliated masses, or powdery incrustations.

Specific Gravity. 3.4 - 3.5.

Hardness: $1\frac{1}{2}$ - 2.

Streak: pale lemon-yellow.

Distinguishing Features: yellow color.

Occurrence and Association: a vein mineral usually associated with realgar.

Similarities: none.

Uses: synthetic arsenic trisulfid, known as "king's yellow" is used as a pigment.

ORTHOCLASE
(or'-tho-klaze)
(see also Feldspar and Plagioclase)

Orthoclase is the most common member of a group of feldspars called potassium, or alkali, feldspars on account of their composition. Orthoclase is found in the more acidic or siliceous rocks.

Color: white, grey, pink or red.

Luster: vitreous to pearly. The succession of thin cleavage planes that give rise to the step-like appearance provide the conditions necessary for pearly luster.

Cleavage: in two directions at right angles. The cleavages may have a step-like appearance which results from the break across tiny cleavage surfaces, and may be best observed by holding the rock or specimen toward the light and turning it in various directions Light is reflected from these surfaces as from a mirror. It may be noticed that only half the crystal reflects light, that a slight turn or rotation will reflect light from the other half as the first goes dark. This feature is a result of what is known as Carlsbad twinning.

Fracture: In some white varieties the fracture may be conchoidal.

Form: orthoclase occurs disseminated through rock masses and as attached and embedded crystals.

Specific Gravity: 2.54 to 2.6.

Hardness: 6 to 6.5.

Streak: white.

Distinguishing Features: orthoclase is difficult to distinguish from plagioclase. If an example of Carlsbad twinning can be found, the feldspar is probably orthoclase while polysynthetic twinning is certain to be developed in plagioclase. . . . . Often it is necessary to identify it merely as feldspar.

Occurrence: orthoclase is usually found in the more siliceous rocks, such as granite, diorite, etc., but may also occur in a certain few basic rocks. The arkosic sandstones of western Oregon carry some orthoclase. Large masses of orthoclase are found in pegmatites, and in this form the mineral has an economic value in pottery and china clay operations. Feldspars alter to kaolin, which is a type of "pure" china clay. Slight alteration may cloud some cleavage surfaces. Complete alteration may leave clay (kaolin) in place of the feldspar, or it may be removed entirely and deposited elsewhere.

Similarities: plagioclase.

Uses: ceramics.

PHOSPHATE ROCK

Phosphate rock is composed largely of the amorphous (non-crystalline) mineral collophane, a calcium carbonate-phosphate. It is a difficult material to identify without blowpipe or chemical tests because of its indefinite character. Suspected phosphate rock should be sent to a laboratory for identification if means are not available for making the test in the field.

PITCHBLENDE

Pitchblende is a complex compound of uranium and oxygen having no definite chemical composition. Radium is present in very small varying amounts.

Color: grey-green to black.

Luster: pitchy or greasy to dull.

Cleavage: none.

Fracture: conchoidal to uneven; brittle.

Form: massive.

Specific Gravity: 6.5 to 8.0.

Hardness: $5\frac{1}{2}$

Streak: olive-green to black.

Distinguishing Features: its high specific gravity distinguishes it from black
     minerals of pitchy luster.   If placed on an unexposed photographic plate it
     will fog the film.

Similarities: it is similar to pitchstone or obsidian, in color, fracture, and
     luster, but it is easily determined by its specific gravity.

Occurrence: in metalliferous veins with sulphides of lead, iron, copper, cobalt
     and zinc.

Uses: it is a source of uranium and radium.

PLAGIOCLASE
(plaj'-ee-oh-klaze)
(see also Feldspar and Orthoclase)

     Plagioclase is one of the principal rock forming minerals. It belongs to the
feldspar group, and is actually a group of minerals itself. (See Feldspar). The
plagioclase minerals vary in composition from sodium (albite) to calcium (anorthite) aluminum silicate, and are also known as the soda-lime feldspars.

Color: white to grey, sometimes with a play of colors as in labradorite.

Luster: vitreous.

Cleavage: perfect in one direction and less perfect in another at nearly right
     angles to the first.   Grooves or striations that are parallel to the long
     edge of the cleavages are a sure indication that the feldspar is a plagioclase.

Fracture: poorly developed.

Form: it seldom occurs in crystals which can be identified as such.   As a rule,
     in basic rocks plagioclase has a lath-shape.

Specific Gravity: 2.6 - 2.7.

Hardness: 6.

Streak: white.

Distinguishing Features: similar to orthoclase.

Occurrence: Plagioclase is an essential feldspar in the more basic igneous rocks,
     particularly andesites and basalts.   Some plagioclase (usually albite) oc-
     curs in the siliceous igneous rocks.   If found in sedimentary rocks, it has
     resulted from the decomposition of an igneous rock containing plagioclase.
     Also it is contained in some metamorphic rocks.   In fact, all rocks should
     be suspected of containing some feldspar until proven otherwise. Alteration:
     to saussurite, a greenish aggregate of zoisite minerals,also to clays,similar
Similarities: orthoclase.   to orthoclase.

Uses: similar to orthoclase.

PLATINUM

Platinum is a heavy, silvery white metal found in rounded grains, scales, irregular lumps and sometimes as nuggets. If iron is alloyed with platinum it may be slightly magnetic.

Color: silvery white.

Luster: bright metallic.

Cleavage: none.

Fracture: hackly; malleable.

Form: rounded grains, scales, irregular lumps and sometimes nuggets.

Specific Gravity: 14-19.

Hardness: $4\frac{1}{2}$.

Streak: light steel grey.

Distinguishing Features: high specific gravity and malleability.

Occurrence. found in placers and in basic (non-siliceous) rocks such as perid-
        otites and dunites with chromite, olivine and serpentine.

Similarities: slightly-rusted to bright iron or steel shavings.

Uses: jewelry; scientific uses; catalyst.

PROUSTITE (Ruby Silver)
(proos'-tite)

Proustite is one of the so-called "ruby silver" ores and is chemically classed as a silver-arsenic-sulfide (Ag_3AsS_3). Pyrargyrite, a silver-antimony-sulfide (Ag_3SbS_3) is the other "ruby silver". Color and streak may distinguish the two.

Color: scarlet-vermilion.

Luster· adamantine.

Cleavage: one direction, good.

Fracture: conchoidal to uneven.

Form: massive, and as distinct crystals.

Specific Gravity: 5.57 - 5.64.

Hardness: 2 - 2½.

Streak: Scarlet-vermilion.

Distinguishing Features: it has a lighter color and streak than pyrargyrite. Its similarity to cuprite and cinnabar requires blowpipe tests to distinguish between them.

Occurrence: as a vein mineral .

Similarities: pyrargyrite, cuprite, and cinnabar.

Uses: ore of silver.

PSILOMELANE (Black Hematite)
(sil-om'-e-lane)

Psilomelane is an impure hydrous oxide of manganese.

Color: black to brown.

Luster: earthy, submetallic to dull, opaque.

Cleavage: none.

Fracture: conchoidal, if any.

Form: amorphous, no crystalline structure. It is a compact, massive to earthy mineral. Sometimes in botryoidal and stalactitic shapes.

Specific Gravity: 3.7 to 4.7, depending on its purity.

Hardness: 5-6, depending on its purity.

Streak: brownish black to brown.

Distinguishing Features: its hardness distinguishes it from pyrolusite and manganite, and its darker streak from limonite.

Occurrence: occurs in residual clays and in bog deposits. It may occur as an alteration product with other manganese minerals.

Similarities: pyrolusite, manganite, limonite.

Uses: ore of manganese.

PUMICE AND PUMICITE.

Pumice and pumicite are not minerals, but rocks, consisting of a fairly pure aggregate of a mineral compound. They are composed mainly of fine vesicular, fibrous glass. Their chemical analyses show the composition to be very similar to the more siliceous lavas.   Often pumice is porphyritic, containing crystals of feldspars, amphiboles and pyroxenes.   Both result from the volcanic eruption of salic lavas.

Pumice is very porous and light of weight.   It is found in fragments ranging in size from sand to large blocks.   Because of the abundance of sealed glass cells some specimens will float indefinitely.

Pumicite is volcanic ash or dust which has resulted from thecomplete disruption of pumice during a volcanic explosion.   The winds spread the dust outward from the source over the surrounding country, upon which it usually settles as a thin blanket   Sometimes thick deposits have been formed where the wind had drifted the ash into a protected canyon or by the streams having transported and deposited it in a lake.

Because pumice and pumicite are chemically and physically nearly the same they will be described as one.

Color   light colored, mostly white and grey.

Luster   vitreous to dull; separate grains are glassy appearing.

Form.   soft porous beds, which form powdery talus slopes.

Specific Gravity. around 2.5 when all the air-cells are broken down.

Hardness: 5.5 to 6

Distinguishing Features: glassy appearance, porous texture and grittiness, when crushed between the teeth.   Insoluble in acids.

Occurences: in stratified layers interbedded with sedimentary and volcanic rocks.   Usually abundant in a region of fairly recent volcanic activity.

Similarities   diatomite, silty clay.

Uses: about 88 percent is used for cleansing and scouring compounds, and the rest for abrasives, cement admixtures, acoustic stucco and plaster.

PYRARGYRITE   (Ruby Silver)
(pie-rar'-gee-rite)

Pyrargyrite is one of the so-called "ruby silver" ores, the other being proustite.   The principal differences between the two are color and streak. Pyrargyrite is a silver antimony sulfide (Ag_3SbS_3) and proustite is silver-arsenic-sulfide (Ag_3AsS_3)

Color: black to dark red; thin plates are deep red.

Luster: metallic-adamantine.

Cleavage: one direction good; one direction poor.

Fracture: conchoidal to uneven.

Form: massive. Crystals are prismatic.

Specific Gravity: 5.7 - 5.9.

Hardness: $2\frac{1}{2}$.

Streak: purplish red.

Distinguishing Features: red fragments and purplish red streak - proustite has
     a brighter color and streak.

Occurrence: a vein mineral formed by ascending solutions.

Similarities: proustite; cuprite, and cinnabar.

Uses: ore of silver.

PYRITE (Fool's Gold)
(pie'-rite)

Pyrite is iron sulfide, a compound of iron and sulphur (FeS_2).

Color: silvery to pale brass-yellow and may tarnish brown with a beautiful play
     of colors (iridescence).

Luster: metallic.

Cleavage: imperfect, cubic.

Fracture: conchoidal to uneven.

Form: pyrite crystallizes in the isometric system and is often found as cubes,
     pyritohedrons, octahedrons, and as massive and granular bodies. Pyritohedron
     faces have 5 edges.   The crystal faces are often striated, or grooved.

Specific Gravity: 5.0.

Hardness: 6 to 6.5.

Streak: greenish to brownish black.

Distinguishing Features: the hardness and brittleness of pyrite distinguish it
     from gold.   Chalcopyrite and pyrrhotite can be scratched with a knife, where-
     as pyrite can not.   It is distinguished from marcasite with difficulty.

Occurrence: pyrite is the most common and widely distributed of the sulfides.
     It may occur in a rock as a result of the primary, or original deposition
     from a molten condition, or it may be secondary.   Solutions containing iron
     are frequently reduced by carbon and we often find pyrite in coal and in car-
     bonaceous sediments.   Waters carrying soluble sulfides may also deposit py-
     rite from solution.

Alteration: pyrite tends to alter to limonite, the "iron-rust" mineral. Many times the alteration has proceeded in a manner whereby the limonite takes the crystal form of the pyrite (pseudomorphism) by replacement. Each atom of oxygen replaces an atom of sulphur and we say that the limonite is a pseudomorph after pyrite.

Similarities: marcasite, pyrrhotite, chalcopyrite, gold, and arsenopyrite.

Uses: pyrite is not classed as an iron ore, although the pure mineral contains 46.6% iron. It is used in the manufacture of sulfuric acid, but sulfur is replacing pyrite for that use. Gold is sometimes associated with pyrite, but its presence cannot be predicted by an "eye-ball assay".

PYROLUSITE (Black Oxide of Manganese)
(pie'-row-lue'-site)

Pyrolusite is a hydrous manganese oxide, $MnO_2(H_2O)_x$.

Color: black.

Luster: metallic to dull; opaque.

Cleavage: none.

Fracture: splintery; often distinguishable.

Form: fibrous and columnar; massive.

Specific Gravity: 4.8.

Hardness: $2-2\frac{1}{2}$; soils the fingers very readily.

Streak: black.

Distinguishing Features: its softness distinguishes it from the other black manganese minerals.

Occurrence: usually an alteration product of manganite.

Similarities: manganite, psilomelane, limonite.

Uses: ore of manganese.

PYROXENE
(pie'-rox-ene)

Pyroxene is a general name applied to a group of complex silicate minerals, of which the principal ones are:

augite   rhodonite (which see)
diopside   hedenbergite
hypersthene   diallage
aegirite (acmite)   spodumene
enstatite (and bronzite)

It is very difficult to distinguish some of these minerals by megascopic
(hand lens) methods, so they are discussed as a group.

Color: usually black, but may be grey, green, brown, and even white (diopside),
    and pink (rhodonite).

Luster: vitreous to pearly.

Cleavage: two well-developed cleavages, parallel to the length of the crystals;
    at nearly right angles to each other.

Fracture: uneven; brittle.

Form: the crystals are stubby as a rule, and have a cross section that is square,
    or nearly so. They sometimes occur in bladed and columnar masses.

Specific Gravity: 3.2 to 3.6.

Hardness: 5 to 6.

Streak: white to grey-green.

Distinguishing Feature: (see amphibole)

	Pyroxene	Amphibole	Epidote	Tourmaline
Color	dark brown, green, black; may be red or white.	dark green, black; may be light green, white.	usually a distinctive pistachio-green	black; also pink, pale green.
Cleavage	two cleavages, with angle between planes 87° & 93° ▢	two cleavages, with angle between planes 56° & 124° ▱	one cleavage	poor cleavage
Form	short crystals	long crystals, sometimes fibrous	long crystals, with striations	long crystals, with striations
Cross-section of crystal	often same as cleavage, nearly square	six sided	six sided	triangular △
Hardness	5-6	5-6	6-7	7.7-5

Occurrence: pyroxenes are associated with igneous rocks, especially the basic
    ones.   Pyroxenes are characteristic of basalts and gabbros, and do not com-
    monly occur in rocks that contain quartz. They are readily decomposed on weath-
    ering and are not usually found in sedimentary rocks.

Alteration: pyroxenes readily alter to carbonates and oxides and are generally
    the first minerals in basic igneous rocks to be affected by weathering. Un-
    der intense metamorphism they may alter to amphibole.

Similarities: amphibole, epidote, tourmaline.

Uses: . . . . Jadeite (included in the pyrox-
    ene group) is used as a semi-precious and ornamental stone, especially in
    China.

PYRRHOTITE (see p. 98b)

QUARTZ
(kwarts)

Quartz is one of the most common minerals.   Many valuable minerals are associated with siliceous (acid) rocks of which quartz is an important component.
Chemically it is composed of silicon dioxide, SiO_2, of which there are three main
varities:

1.   Quartz
2.   Chalcedony (see chalcedony)
3.   Opal   (see opal)

Color: colorless, white, rose, amethyst, etc.

Luster: vitreous.

Cleavage: none.

Fracture: conchoidal.

Form: crystals of hexagonal form with pointed pyramidal ends are common in places
    where they have had an opportunity to grow unhampered.   More frequently quartz
    is quite massive, the crystalline structure not being apparent to the eye.

Specific Gravity: 2.6 - 2.7.

Hardness 7.

Streak: none.

Distinguishing Features: the hardness and lack of cleavage are quite distinctive.
    Usually has a glassy appearance.

Occurrence: found in most acid igneous rocks; almost never found in basic rocks.
    Found as grains in sandstones; as massive and crystalline quartz in veins,
    formed by ascending or descending solutions.

Similarities: beryl, scheelite, white feldspars.

Uses: crystals of large size are used in the manufacture of optical lenses.
     Massive varieties are used as flux, refractory, abrasive, grits, etc. Near-
     ly pure quartz sands and sandstones are used in the manufacture of glass.

REALGAR
(ree-al'-gar)

     Realgar is an arsenic sulfide, occurring as an incrustation or as dissemin-
ated grains.

Color: red, which changes to orange as it alters to orpiment.

Luster: adamantine to resinous.

Cleavage: generally not distinguishable.

Fracture: conchoidal.

Form: massive and granular crystals.

Specific Gravity: 3.5.

Hardness: $1\frac{1}{2}$ to 2.

Streak: orange yellow.

Distinguishing Features: lower specific gravity and hardness than cinnabar.

Occurrence: vein mineral associated with stibnite, cinnabar, pyrite, etc.

Similarities: cinnabar.

Uses: minor source of arsenic.

RHODOCHROSITE
(roe'-doe-crow'-site)

     Rhodochrosite is manganese carbonate ($MnCO_3$).

Color: pink to brownish red; also yellowish grey and tan.

Luster: vitreous to pearly.

Cleavage: in 3 directions at oblique angles - rhombohedral, - like calcite.

Fracture: uneven; brittle.

Form: in rhombohedral masses, and as crystals; occasionally colloform.

Specific Gravity: 3.5 - 3.6.

Hardness: $3\frac{1}{2}$ - $4\frac{1}{2}$.

Streak: white.

Distinguishing Features: rhodochrosite is heavier than calcite or dolomite, and is softer than rhodonite; soluble with effervescence in warm hydrochloric acid.

Occurrence: in veins with silver and lead ores; also found with other manganese ores.

Similarities: calcite, dolomite, magnesite, and rhodonite.

Uses: an ore of manganese.

RHODONITE
(row'-doe-nite)

Rhodonite is manganese silicate ($MnSiO_3$) belonging to the pyroxene group, but easily distinguished by its color.

Color: pink to red, sometimes stained black by manganese oxides.

Luster: vitreous.

Cleavage: in two directions at nearly right angles ($92\frac{1}{2}°$).

Fracture: conchoidal to uneven.

Form: cleavable and compact masses. Crystals are triclinic.

Specific Gravity: 3.4 - 3.6.

Hardness: $5\frac{1}{2}$ - $6\frac{1}{2}$; tough.

Streak: white.

Distinguishing Features: its hardness is greater than rhodochrosite, and its cleavage is more rectangular than rhombohedral. It has higher specific gravity than orthoclase.

Occurrence: rhodonite occurs in veins and as accessory mineral in various deposits.

Similarities: rhodochrosite, magnesite, orthoclase.

Uses: rhodonite is used as an ornamental stone and as a gem material.

RUTILE
(rue'-teel)

Rutile is a titanium oxide (TiO_2).

Color: brownish red to black.

Luster: adamantine to sub-metallic.

Cleavage: unimportant.   It is imperfect, at right angle.

Fracture: uneven; brittle.

Form: embedded grains or crystals.   Long, thin, acicular crystals are found penetrating other crystals.

Specific Gravity: 4.2.

Hardness: 6-6$\frac{1}{2}$.

Streak: pale brown.

Distinguishing Features: its specific gravity is lower than cassiterite; crystallization, color and adamantine luster are distinctive.

Occurrence: as accessory mineral in veins, and in pegmatites; commonly as crystals embedded in quartz or feldspar; as secondary mineral; and as constituent of auriferous sands.

Similarities: cassiterite, zircon.

Uses: coloring material for porcelain; manufacture of ferrotitanium; paints.

SCHEELITE
(she'-lite)

Scheelite is calcium tungstate ($CaWO_4$), one of the important ores of tungsten, and is important as a strategic war mineral.

The mineral is rather difficult to determine by ordinary field methods, but its high specific gravity allows it to be detected even when disseminated throughout the rock, by ordinary panning, in the same manner as gold.   Its appearance is very similar to that of quartz, but it may be positively identified by the use of the ultra-violet lamp.

Color: white, grey, or pale colors.

Luster: sub-adamantine.

Cleavage: in one direction, interrupted.

Fracture: uneven, brittle.

Form: crystals are tetragonal, while massive forms show little crystal development.

Specific Gravity: 6.0.

Hardness: $4\frac{1}{2}$ to 5.

Streak: white.

Distinguishing Features: its high specific gravity and luster; its pale-blue fluorescence in ultra-violet light is distinctive.

Occurrence: in veins in crystalline rocks, associated with cassiterite, topaz, apatite and wolframite in quartz. In Oregon found at contacts of granite and limestone, associated with garnet, epidote, molybdenite, etc.

Similarities: quartz.

Uses: ore of tungsten, which is used widely in manufacture of tool steels.

SERPENTINE
(ser'-pen-tine)

Serpentine is considered to be a hydrated magnesium silicate, with formula $H_4Mg_3Si_2O_9$ or $3MgO.2SiO_2.2H_2O$. It usually contains a small amount of iron replacing magnesium.

Color: usually greenish; also brownish, reddish, and variegated.

Luster: sub-resinous to greasy.

Cleavage: not important.

Fracture: conchoidal to splintery.

Form: usually massive; sometimes fibrous.

Specific Gravity: 2.5 - 2.65.

Hardness: variable, $2\frac{1}{2}$-4.

Streak: white, slightly shining.

Distinguishing Features: softness, greasy luster, "feel", and absence of cleavage.

Occurrence: commonly as masses, often of great size, resulting from alteration of certain rocks, notably peridotite. The fibrous variety, chrysotile, is described under asbestos.

Similarities: talc, into which it may grade; chlorite.

Uses: ornamental stone (verde antique); mineral filler.

SIDERITE
(sid'-er-ite)

Siderite is iron carbonate ($FeCO_3$) and is a minor ore of iron.

Color: brown to grey.

Luster: highly vitreous to sub-adamantine

Cleavage: 3 directions at oblique angles (rhombohedral), like that of calcite, with curved crystal faces like dolomite.

Fracture   uneven; brittle.

Form: siderite occurs in crystals much like those of calcite. Some of the crystal faces may be curved, like those on dolomite   It more usually is found 'n cleavable masses, crusts, concretions, and oolites.

Spec'fic Gravity: 3.8.

Hardness: 3.5 to 4.

Streak: pale brown to grey.

Di.tinguishing Features: it is similar to 'alcite and dolomite, having the same cleavage, but it has a higher specific gravity   Its color and general ap pearance is like that of sphalerite, but its rhombohedral cleavage and high luster is distinct've.

Occurrence: Siderite is usually a gangue mineral in deposits that are formed by replacements.   A variety known as spherosiderite occurs in cavities in basalt and was first recognized by amateur mineralogists at Spokane.

Similarities: calcite, dolomite, sphalerite.

Uses: it is a minor ore of iron.

SILVER

The more important silver minerals are:

*Native Silver (Ag)	Proustite Ag_3AsS_3
*Argentite Ag_2S	Stephan'te Ag_5SbS_4
Sylvanite $AuAgTe_4$	Polybasite$(Ag.Cu)_{16}Sb_2S$
Pyargyrite Ag_3SbS_3	*Cerargyrite $AgCl$

Those starred are described in this bulletin.

Native silver is found at a number of localities, sometimes in large masses. It may contain up to 10 percent gold, or small amounts of copper platinum, antimony, bismuth, mercury.

Color: tin white to pale yellow - usually dull and tarnished.

Luster: metallic.

Cleavage: none

Fracture: hackly; ductile and malleable.

Form: wire, thin sheets, skeleton crystals, rarely as nuggets.

Specific Gravity: 10.1 - 11.1.

Hardness: $2\frac{1}{2}$-3.

Streak: silver-white, metallic.

Distinguishing Features: color of freshly cut surface, and malleability.

Occurrence: a vein mineral associated with cobalt, nickel, copper, and lead; and with other silver minerals from which it was derived. In the gossan, or "iron hat" of some ore deposits, it occurs as native silver. It is rarely found as nuggets, in placer deposits.

Similarities: galena, native antimony, and stibnite.

Uses: monetary; in the arts; and in manufacture of chemicals.

SMALTITE (Co,Ni) As$_2$
(small'-tite)

Smaltite is considered to be a diarsenide of cobalt, but nickel is usually present. The mineral has been found in the ores of certain Oregon mines.

Color: tin white to steel grey, sometimes irridescent (rainbow colors).

Luster: metallic.

Cleavage: absent.

Fracture: granular; brittle.

Form: massive, without cleavage, although cubic crystals may be found occasionally.

Specific Gravity: 5.5 - 6.8.

Hardness: $5\frac{1}{2}$-6.

Streak: greyish-black.

Distinguishing Features: color, specific gravity. Smaltite is very similar to arsenopyrite and to cobaltite but may be distinguished from them by chemical and microscopical tests.

Occurrence: it occurs with silver and copper ores and in veins with ores of nickel and copper. Oxidation of smaltite results in erythrite, a pink, earthy mineral, known as "cobalt bloom" which characterizes outcrops of some deposits of cobalt minerals.

Similarities: arsenopyrite; cobaltite; pyrite.

Uses: ore of cobalt.

SMITHSONITE (Dry-bone)
(smith'-son-ite)

Smithsonite is zinc carbonate, $ZnCO_3$.

Color: white to varied; sometimes blue or green.

Luster: vitreous.

Cleavage: has 3 directions of cleavage at oblique angles (rhombohedral) with curved edge like dolomite, but they are not prominently developed.

Fracture: uneven; brittle.

Form: it occurs as nodular masses and rough (drusy) crusts. Rarely found in crystal form.

Specific Gravity: 4.3 - 4.5.

Hardness: 5.

Streak: colorless.

Distinguishing Features: harder than other carbonates that have rhombohedral cleavage. The green and blue varieties resemble chrysocolla but are heavier and harder.

Occurrence: smithsonite occurs in the oxidized zone as an alteration product of sphalerite.

Similarities: chrysocolla, calcite, dolomite.

Uses: minor ore of zinc.

SPHALERITE (Rosin Jack, Black Jack, Zinc Blende)
(sfal'-er-ite)

Sphalerite is an important ore of zinc and its name means "treacherous" indicating that it is difficult to determine. The Germans call it blende, meaning deceptive. It is zinc sulfide (ZnS).

Color: variable, pale yellow (rosin jack) to black (black jack) depending on amount of iron present. Rarely colorless. Some of it is triboluminescent (develops an interior glow with friction).

Luster: resinous to adamantine.

Cleavage: is very prominent. Planes intersect at angles of 60°, a type known as dodecahedral.

Fracture: conchoidal; brittle.

Form: massive forms are most common; . . . . isometric, (like a cube), or some of its modifications. Some of the faces may be striated (grooved).

Specific Gravity: 4.0.
Hardness: 3.5 - 4.
Streak: pale yellow to dark brown.

Distinguishing Features: sphalerite looks like a number of other minerals. Its perfect cleavage and resinous luster are distinguishing features. It is much softer than garnet.

Occurrence: associated with galena, chalcopyrite, pyrite, and other sulfides in metalliferous veins. Often found in limestones and in contact-metamorphic zones.

Similarities: garnet, siderite.

Uses: ore of zinc.

STIBNITE
(stib'-nite)

Stibnite is antimony sulfide, Sb_2S_3, and, when pure, contains 71.4 percent antimony.

Color: lead grey.

Luster: brilliant metallic on fresh surfaces, but on exposed faces a lead-grey tarnish develops.

Cleavage: perfect in one direction; parting is sometimes evidenced by lines on the cleavage faces.

Fracture: uneven.

Form: prismatic or acicular crystals, bladed aggregates, or granular masses.

Specific Gravity: 4.5.

Hardness: 2, (it can be scratched with the finger nail).

Streak: lead grey.

Distinguishing Features: crystal form, hardness, low fusibility (it will soften in a match flame) and perfect cleavage are distinguishing features.

Occurrence: vein mineral in quartz with zinc, ^lead, and mercury sulphides; sometimes associated with metallic gold.

Similarities: galena; rutile.

Uses: important source of antimony, which is used in storage battery lead; in babbitt metal, in type metal, and in manufacture of chemicals and munitions.

SULFUR (Brimstone)

    Sulfur (S) is well known and generally easily determined.

Color: yellow to yellowish-orange and-red, occasionally grey.

Luster: resinous to adamantine.

Cleavage: absent.

Fracture: conchoidal, very brittle.

Form: crystals; crusts; masses.

Specific Gravity: 2.

Hardness: $1\frac{1}{2}$-$2\frac{1}{2}$; varying with amount of impurities.

Streak: colorless.

Distinguishing Features: yellow color, low hardness, brittleness, odor of sulfur dioxide when burned.

Occurrence: its association with gypsum as a result of volcanic activity is well known; results from decomposition of pyrite in coal beds, where it occurs in hot springs deposits, and in regions of active or extinct volcanoes. Combined with other elements, sulfur forms sulphides and sulphates, which occur widely distributed in mineral deposits.

Similarities: none.

Uses: in the manufacture of sulfuric acid; in paper making; in manufacture of gunpowder, fireworks, insecticides, fertilizers, etc.

TALC (Steatite, Soapstone)

Talc is a secondary mineral formed as the result of rock alteration. It is commonly associated with serpentine, and with talcose or chloritic schist. Chemically it is hydrous magnesium silicate and may contain small quantities of iron and aluminum. It is very resistant to heat and may be used directly as refractory material.

Color: white, grey, or pale green.

Luster: pearly.

Cleavage: perfect in one direction   ; compact varieties show no cleavage, as a rule.

Fracture: none. Flexible in thin laminae, but not elastic.

Form: compact, massive, in scales, foliated (like matted leaves), and fibrous masses.

Specific Gravity: 2.7.

Hardness: when pure, 1.

Streak: white.

Distinguishing Features: foliated masses may be mistaken for mica but the sheets are not elastic.   Talc has a "soapy" feel.

Occurrence: widely distributed and results from metamorphism (alteration by heat or pressure) from basic rocks.   In many cases metamorphic action produces a rock called schist and if sufficient talc is present, it may be classed as a talcose schist.   Purity varies over a wide range.

Similarities: mica; chlorite.

Uses: talcum powder; refractories; glaze for paper stock; mineral filler.

TELLURIDES
(tell'-your-ides)

Tellurides are compounds of tellurium with other metals, usually gold and silver.   They are not common, and are difficult to determine in the field; blowpipe, chemical, or optical methods are usually necessary in order to determine them definitely.

Metallic tellurium (Te) is rare, but has been found in some Colorado and European mines, associated with tellurides.   The best known tellurides are:

Sylvanite	$(Au\ Ag)Te_2$
Calaverite	$AuTe_2$
Hessite	Ag_2Te
Petzite	$(Ag\ Au)_2Te$

TETRAHEDRITE (Grey Copper)
(tet'-rah-he'-drite)

Tetrahedrite, frequently called "grey copper", is a compound of copper, antimony and sulfur, with varying amounts of iron, zinc, and sometimes silver as impurities. The pure mineral contains 52% copper; the silver-bearing variety may contain up to 30% silver.

Color: dark iron grey.

Luster: metallic.

Cleavage: none.

Fracture: uneven; brittle.

Form: its name suggests that it occurs as tetrahedrons, a four-sided figure, with 3 edges on each side. . . . . The usual form, however, is massive.

Specific Gravity: 4.4 - 5.1.

Hardness: 3-4$\frac{1}{2}$.

Streak: dark iron grey.

Distinguishing Features: When crystallized into tetrahedrons, the crystal form and color are distinguishing characteristics. Massive tetrahedrite can be mistaken for chalcocite, but it is brittle and chalcocite is sectile.

Occurrence: It occurs as a vein mineral with chalcopyrite, galena, sphalerite, and siderite. The massive ore is not uncommon in copper and silver mines.

Similarities: chalcocite.

Uses: ore of copper, and occasionally silver.

TITANIUM MINERALS
(tie-tane'-ee-um)

The most important titanium minerals are described under the names of ilmenite, rutile, and titanite.

TITANITE (Sphene)
(tie'-tan-ite)

Titanite is calcium-titanium silicate, $CaTiSiO_5$.

Color: varying tints of yellow or brown.

Luster: adamantine to high vitreous.

Cleavage: indistinct; brittle.

Fracture: none.

Form: Flat, wedge-shaped crystals; massive; granular.

Specific Gravity: 3.4 - 3.5.

Hardness: 5-5½.

Streak: none, or white.

Distinguishing Features: crystal shape.

Occurrence: it occurs in granitoid rocks with magnetite and hornblende, as an
      accessory mineral.   In metamorphic rocks, it probably occurs as a decompo-
      sition product of pyroxenes.

Similarities: magnetite, hornblende.

Uses: perfect crystals have been used as gem stones.

TOURMALINE
(tour'-ma-lene)

Tourmaline is a complex silicate of boron and aluminum containing chemically
combined water, and is found as an accessory mineral in many rocks.   It occurs in
a variety of colors, and is pyro-electric, that is, a crystal which has been heated
will on cooling develop positive electricity at one end and negative electricity at
the other.

Color: usually black, brownish black, bluish black; also blue, green, red, and
      sometimes crystals are red at one end and green, blue, or black at the other.
      Transparent to opaque.

Luster: vitreous to resinous.

Cleavage: none.

Fracture: sub-conchoidal to uneven.

Form: prismatic (long, slender) crystals ranging in size from slender, small
      needles to those that are 4 inches and more in diameter.   Sides of crystals
      show striations.

Specific Gravity: 3 - 3.25.

Hardness: 7 - 7½.

Streak: uncolored.

Distinguishing Features: a crystal cross section is rounded triangular, and it has no cleavage, as does hornblende; crystals often deeply striated.

Occurrence: in pegmatites, granites, and gneisses, and certain other metamorphic rocks.

Similarities: hornblende, pyroxene (which see).

Uses: the transparent red and green varieties are used as semi-precious gem stones, known variously as rubellite, indicolite, Brazilian sapphire, Brazilian emerald, Peridot of Ceylon (peridot is gem quality olivine), achroite.

WOLFRAMITE
(wolf'-ram-ite)

Wolframite is one of the principal tungsten minerals and is an iron, manganese, tungsten oxide $(Fe,Mn)WO_4$.

Color: dark brown to black.

Luster: sub-metallic to metallic.

Cleavage: perfect in one direction.

Fracture: uneven.

Form: occurs in crystals or aggregates of crystals, somewhat tabular in shape.

Specific Gravity: 7.2 - 7.5.

Hardness: 5 to $5\frac{1}{2}$.

Streak: black to dark reddish brown.

Distinctive Features: its high specific gravity, cleavage in one direction, and metallic luster.

Occurrence: wolframite is found in pegmatites and in veins associated with siliceous rocks.

Similarities: hubnerite $(MnWO_4)$ is very similar to wolframite; for general field purposes hubnerite may be considered as brown in color and having a brown streak, while wolframite is generally black in color and gives a black streak.
Uses: Ore of tungsten, which is used in tool steels and other steel alloys.

ZIRCON
(zir'-con)

Zircon, zirconium silicate $(ZrSiO_4)$, is an accessory mineral in acid igneous and metamorphic rocks. It is concentrated in certain sands as it is resistant to abrasion (hardness of $7\frac{1}{2}$) and to chemical action.

Color: reddish-brown or grey; sometimes yellow, red, or colorless, rarely blue
    or green.

Luster: adamantine; brilliant.

Cleavage: imperfect.

Fracture: conchoidal; brittle.

Form: usually found as small grains in river and beach sands, in box-like crys-
    tals with pyramidal ends.

Specific Gravity: averages about 4.7, which is rather high for transparent mat-
    erials.

Hardness: 7.5, harder than quartz.

Streak: uncolored.

Distinguishing Features: Crystal shape and high specific gravity.

Occurrence: loose crystals in sands derived from areas containing granitic and
    syenitic rocks; crystalline rocks especially granular limestone, chloritic
    and other schists; gneiss.

Similarities: quartz, cassiterite, rutile, diamond.

Uses: it is used as a refractory in the form of zirconia bricks; in ferro-alloys,
    and, when pure, as a gem; zirconium compounds are used as welding rod coat-
    ings, and in ceramics; zirconium metal is used in radio tubes.

ZEOLITES
(zee'-oh-lights)

The minerals of this family of hydrous aluminum, calcium and sodium silicates are very similar in composition, association, and mode of occurrence.   Many of them fuse easily, and give off water, hence the name, which in Greek means "boiling stone".   They are secondary after feldspars.

Color· Most commonly colorless to white. Also yellow, reddish, brown, grey.

Luster: Vitreous.

Cleavage: Usually perfect in one direction.

Fracture   Uneven.

Form: Nearly always in twinned crystals, often in sheaf-like aggregates.   Platy, cubes, etc

Specific Gravity: 2-2.4.

Hardness: 3.5 - 5.5.

Streak: Colorless.

Distinguishing Features: Are all softer than quartz, and do not have the cleavage nor effervescence in HCl of calcite.   They usually have good crystal form, but sometimes are in sheafs or in radiating crystals.   Usually light-colored or pinkish.

Occurrence: These minerals are commonly found lining the gas bubbles and cavities in dark colored lavas, as well as in veins in these rocks.

Similarities: Calcite, feldspar, fluorite, garnet, aragonite.

ADDENDA

CORUNDUM
(kor-un'-dumb)

Corundum is aluminum oxide in its purest natural form, Al_2O_3.   Sapphire, ruby, and emery are all varieties of this mineral, whose brilliant luster and extreme hardness make it ideal as a gem stone.

Color: Gray, brown, nearly white, blue (sapphire), red (ruby).

Luster: Adamantine to vitreous.

Cleavage: In one direction sometimes perfect.

Fracture: Uneven to conchoidal.

Form: Crystals usually rough and rounded, six-sided.   Also massive.

Specific Gravity: 4.

Hardness: 9.

Streak: none.

Distinguishing Features: The extreme hardness, only exceeded in nature by the
        diamond, is the most important characteristic. Its adamantine luster and
        high specific gravity are also important.

Occurrence: In metamorphosed rocks and granite, occasionally in other igneous
        rocks, especially those lacking quartz. Burmese rubies are mined from
        placer deposits.

Similarities: Massive variety may resemble cleavable feldspar, but is harder
        and denser.

Uses: Clear varieties form valuable gem stone (above). Also formerly used as
        an abrasive.

PYRRHOTITE

(peer'-ho-tite)

Pyrrhotite or "magnetic pyrites" is an iron sulphide with a variable composition, although there is always one more "S" than there is "Fe" in the formula. It often contains small amounts of nickel. It is the only sulphide
that is magnetic.

Color: Between bronze-yellow and copper-red, but tarnishes easily.

Luster: Metallic.

Cleavage: Usually massive, sometimes cleaves in one direction.

Fracture: Uneven to subconchoidal.

Form: Distinct crystals rare, usually granular massive.

Specific Gravity: 4.6.

Hardness: 3.5 - 4.5.

Streak: Dark greyish black.

Distinguishing Features: It is the only magnetic sulphide mineral; and the pecul-
        iar reddish-bronze color is characteristic.

Occurrence: In veins with other sulphides, in large masses in igneous rocks.

Similarities: Niccolite, pyrite, marcasite, chalcopyrite.

Uses: It is an important ore of nickel at Sudbury, Ontario.

FIGURE 14

MINERALS ARRANGED AS TO SPECIFIC GRAVITY

1-2		
1.1 - 1.7	Coal
1.5 - 2.0	Nitrates
1.7	Borax
1.9 - 2.4	Diatomite

2-3		
1.9 - 2.4	Diatomite
2.0	Sulfur
2.0 - 2.3	Opal
2.0 - 2.2	Chrysocolla
2.1 - 2.6	Halite
2.2	Graphite
2.2 - 3.2	Asbestos
2.3	Gypsum
2.5	Bauxite
2.5	Pumice
2.5 - 2.6	Orthoclase
2.5 - 2.7	Feldspar
2.5 - 2.7	Serpentine
2.5 - 3.1	Mica
2.6	Chert
2.6	Chalcedony
2.6	Clay
2.6	Quartz
2.6 - 2.7	Plagioclase
2.6 - 2.96	Chlorite
2.7	Beryl
2.7	Calcite
2.7	Talc
2.8 - 3.4	Chlorite
2.8 - 3.0	Dolomite
2.9	Anhydrite

3-4		
2.2 - 3.2	Asbestos
2.6 - 3.4	Chlorite
2.7 - 3.1	Mica
3.0	Magnesite
3.0 - 3.2	Fluorite
3.0 - 3.25	Tourmaline
3.0 - 3.5	Amphibole
3.0 - 3.8	Siderite
3.1 - 4.3	Garnet
3.2	Apatite
3.2 - 3.5	Epidote
3.2 - 3.6	Pyroxene
3.3	Olivine
3.4 - 3.5	Calamine
3.4 - 3.5	Orpiment
3.4 - 3.5	Titanite
3.4 - 3.6	Rhodonite
3.5	Diamond
3.5	Realgar
3.4 - 3.6	Rhodochrosite
3.7 - 4.7	Psilomelane
3.7 - 3.9	Azurite
3.6 - 4.0	Limonite
3.8 - 4.0	Malachite

4-5		
3.1 - 4.3	Garnet
3.7 - 4.7	Psilomelane
3.8 - 4.0	Malachite
3.9 - 4.2	Sphalerite
4.1 - 4.3	Chalcopyrite
4.2 - 4.3	Rutile
4.3	Manganite
4.1 - 4.5	Smithsonite
4.3 - 4.6	Chromite
4.4 - 5.1	Tetrahedrite
4.5	Barite
4.5	Stibnite
4.3 - 5.5	Ilmenite
4.6	Covellite
4.7	Zircon
4.7 - 4.8	Molybdenite
4.7 - 5.3	Monazite
4.8	Pyrolusite

5-6		
4.4 - 5.1	Tetrahedrite
4.5 - 5	Ilmenite
5.0 - 5.3	Hematite
5.0	Pyrite
5.0	Bornite
5.1	Magnetite
5.5	Cerargyrite
5.5 - 5.8	Chalcocite
5.5 - 6.8	Smaltite
5.6	Proustite
5.7 - 5.9	Pyrargyrite

6-7		
5.5 - 6.8	Smaltite
6.0	Arsenopyrite
6.0	Cuprite
6.0	Scheelite
6.0 - 6.3	Cobaltite
6.1 - 6.4	Anglesite
6.5 - 6.6	Cerussite
6.5 - 8.0	Pitchblende
6.8 - 7.0	Cassiterite

7-8		
6.8 - 7.0	Cassiterite
7.1 - 7.5	Wolframite
7.3	Argentite
7.3 - 7.8	Niccolite
7.3 - 7.6	Galena
8.0 - 8.2	Cinnabar

8 - over 8		
8.0 - 8.2	Cinnabar
10.1 - 11	Silver
14 - 19	Platinum
15 - 19	Gold

FIGURE 15

MINERALS ARRANGED AS TO HARDNESS

Minerals are arranged alphabetically under each hardness range. Minor variations within each range are not noted. If a mineral has a hardness that extends through more than one range, it is listed in each range; for example, limonite has a hardness of $1-5\frac{1}{2}$: it will be found listed under 1-2, 2-3, 3-4, 4-5, 5-6.

Hardness of 1-2
Bauxite
Bentonite
Cerargyrite
Chlorite
Clay $(1\frac{1}{2}-2\frac{1}{2})$
Coal $(1-1\frac{1}{2})$
Covellite
Diatomite
Graphite
Gypsum
Limonite $(1-5\frac{1}{2})$
Molybdenite
Nitrates
Orpiment
Realgar
Sulfur $(1\frac{1}{2}-2\frac{1}{2})$
Talc (1-4)

Hardness of 2-3
Argentite
Asbestos $(2\frac{1}{2}-5\frac{1}{2})$
Barite $(2\frac{1}{2}-3\frac{1}{2})$
Biotite
Borax
Chalcocite
Chlorite
Chrysocolla (2-4)
Cinnabar
Clay $(1\frac{1}{2}-2\frac{1}{2})$
Coal $(1-2\frac{1}{2})$
Galena
Gold
Gypsum
Halite
Limonite $(1-5\frac{1}{2})$
Mica $(2\frac{1}{2}-3\frac{1}{2})$
Proustite
Pyrargyrite
Pyrolusite
Serpentine $(2\frac{1}{2}-4)$
Silver
Stibnite
Sulfur $(1\frac{1}{2}-2\frac{1}{2})$
Talc (1-4)

Hardness of 3-4
Anglesite
Anhydrite
Azurite
Asbestos $(2\frac{1}{2}-5\frac{1}{2})$
Barite $(2\frac{1}{2}-3\frac{1}{2})$
Bornite
Calcite
Cerussite
Chalcopyrite
Chrysocolla (2-4)
Cuprite
Dolomite
Limonite $(1-5\frac{1}{2})$
Magnesite $(3\frac{1}{2}-4\frac{1}{2})$
Malachite
Mica $(2\frac{1}{2}-3\frac{1}{2})$
Rhodochrosite$(3\frac{1}{2}-4\frac{1}{2})$
Serpentine $(2\frac{1}{2}-4)$
Siderite
Sphalerite
Talc (1-4)
Tetrahedrite$(3-4\frac{1}{2})$

Hardness of 4-5
Asbestos $(2\frac{1}{2}-5\frac{1}{2})$
Fluorite
Calamine
Limonite $(1-5\frac{1}{2})$
Magnesite $(3\frac{1}{2}-4\frac{1}{2})$
Manganite
Platinum
Rhodochrosite$(3\frac{1}{2}-4\frac{1}{2})$
Scheelite
Tetrahedrite$(3-4\frac{1}{2})$

Hardness of 5-6
Amphibole
Apatite
Arsenopyrite
Asbestos $(2\frac{1}{2}-5\frac{1}{2})$
Chromite
Cobaltite
Feldspar $(5\frac{1}{2}-6\frac{1}{2})$
Hematite $(5\frac{1}{2}-6\frac{1}{2})$
Ilmenite
Limonite $(1-5\frac{1}{2})$
Magnetite $(5\frac{1}{2}-6\frac{1}{2})$
Monazite
Niccolite
Opal $(5\frac{1}{2}-6\frac{1}{2})$
Pitchblende
Psilomelane
Pumice
Pyroxene
Rhodonite $(5\frac{1}{2}-6\frac{1}{2})$
Smaltite
Smithsonite
Titanite
Wolframite

Hardness of 6-7
Cassiterite
Epidote
Feldspar $(5\frac{1}{2}-6\frac{1}{2})$
Garnet $(6-7\frac{1}{2})$
Hematite $(5\frac{1}{2}-6\frac{1}{2})$
Magnetite $(5\frac{1}{2}-6\frac{1}{2})$
Olivene
Opal $(5\frac{1}{2}-6\frac{1}{2})$
Orthoclase
Plagioclase
Pyrite
Rhodonite $(5\frac{1}{2}-6\frac{1}{2})$
Rutile

Hardness of 7-8
Beryl
Chalcedony
Chert
Garnet $(6\frac{1}{2}-7\frac{1}{2})$
Quartz
Tourmaline
Zircon

Hardness of 8-9
Topaz
Spinel

Hardness of 9-10
Corundum (ruby, sapphire)

Hardness of 10
Diamond

FIGURE 16

MINERALS ARRANGED AS TO STREAK

Black:
Chalcocite
Coal
Graphite
Ilmenite
Magnetite
Manganite
Molybdenite (greenish)
Niccolite (brownish)
Pitchblende
Pyrolusite
Wolframite

Brown:
Cassiterite (pale brown)
Chromite (dark)
Coal
Cuprite (brownish red)
Hematite (brownish red)
Ilmenite (brownish black)
Limonite (yellowish brown)
Manganite (reddish)
Niccolite (pale brownish black)
Psilomelane (brownish black)
Rutile (pale brown)
Siderite (pale)
Sphalerite
Wolframite (reddish brown)

Greenish-black:
Chalcopyrite
Molybdenite
Pyrite

Gray-black: (lead grey)
Argentite
Arsenopyrite
Bornite
Chalcocite
Cobaltite
Covellite
Galena
Graphite
Platinum (steel grey)
Smaltite
Stibnite
Tetrahedrite

Blue:
Azurite

Red:
Cinnabar
Cuprite
Proustite
Pyrargyrite
Realgar (orange)

Yellow:
Epidote
Gold
Orpiment
Realgar (orange)
Sphalerite

Green:
Chlorite (grayish)
Malachite (pale)
Pitchblende (olive-green)

Metallic:
Silver

Colorless or white to pale colors:

Amphibole	Fluorite	Sulphur
Anglesite	Garnet	Talc
Anhydrite	Gypsum	Titanite
Apatite	Halite	Tourmaline
Asbestos	Magnesite	Zircon
Barite	Mica	
Bauxite	Monazite	
Bentonite	Nitrates	
Bleaching Clay	Olivene	
Calamine	Opal	
Calcite	Orthoclase	
Cerargyrite	Plagioclase	
Cerussite	Pumice	
Chalcedony	Pyroxene	
Chert	Quartz	
Chrysocolla	Rhodochrosite	
Diatomite	Rhodonite	
Dolomite	Scheelite	
Feldspar	Serpentine	
	Smithsonite	

Name	Color	Luster	Cleavage	Fracture	Form	Sp. Gr.
AMPHIBOLE Complex silicate	Grey to black; varied.	Vitreous to glassy	2 directions making angles between faces of 55° & 125°.	Jagged blades on cleavage edges	Slender crystals, columnar, fibrous, striations parallel to length.	3-3.5
ANGLESITE $PbSO_4$	White to dark grey	Adamantine to dull	Imperfect and not important	Conchoidal; brittle	Crystals; banded masses	6.3
ANHYDRITE $CaSO_4$	White; tinted red and blue.	Pearly to glassy; massive variety is dull.	3 directions at right angles	Uneven in massive varieties	Granular masses of sugar-like texture.	2.9
APATITE $Ca_{10}F_2(PO_4)_6$	White; green; & reddish brown.	Vitreous to sub-resinous	Imperfect	Uneven; brittle	Hexagonal crystals & compact masses	3.2
ARGENTITE Ag_2S	Dark lead grey to dull black.	Metallic	Unimportant; sometimes forming cubes.	Uneven to hackly.	Massive crusts & rough cube-like crystals.	7.3
ARSENOPYRITE FeAsS	Silvery white to steel grey	Metallic	Not important	Uneven	Twin crystals; granular compact masses; columnar forms.	6.
ASBESTOS Complex silicate	Grey brown & green; varied	Silky to dull	Some varieties perfect; acicular	Splintery	Slender, acicular crystals and in fibrous masses.	2.2-3.2
AZURITE $Cu_3(OH)_2 \cdot (CO_3)_2$	Azure-blue	Vitreous; velvety	Not important	Conchoidal	Acicular crystals; nodular groups of crystals; and as crystalline coatings.	3.8
BARITE $BaSO_4$	White & various tints of brown, blue, green & yellow	Vitreous	Perfect in 2 directions; imperfect in the 3rd; 2 cleavages at right angles; 3rd oblique.	Uneven	Tabular and platy crystals; globular & granular masses.	4.5
BAUXITE $Al_2O_3 \cdot 2H_2O$	Shades of brown, yellow and red.	Earthy	None	Earthy to conchoidal	In clay-like masses and small rounded concretions called pisolites	2.5

Hardness	Streak	Distinguishing Features	Occurrence	Similarities	Uses
5-6	White	False hex.cross-section;striations; cleavage; and long-bladed prisms.	Salic igneous rocks;gneisses; mica schists.	Pyroxene; Tourmaline; Epidote.	Fibrous forms as asbestos.
3	White to grey	Does not effervesce in acid; heavier than barite; crystal structure.	Secondary mineral found in oxidized zone of lead deposits.	Cerussite; Barite	Minor ore of lead
3-3½	White	Harder and higher sp.gr. than gypsum; false cubic cleavage is distinct.	In bedded sedimentary deposits; in cave deposits; and as a metamorphic rock.	Gypsum	Fertilizer
5	Colorless	Softer than beryl; about same hardness as glass; and sub-resinous luster.	Salic igneous rocks; metamorphosed rocks; high temp.deposits	Beryl	Used in prep. of superphosphate for fertilizer.
2-2½ Sectile	Shining dark lead grey	Low hardness; high specific gravity & sectility; it is not malleable.	Vein mineral with other silver, lead, and copper sulfides.	Resembles tarnished silver and cuts like lead.	Ore of silver
5½-6; tough.	Greyish-black	Garlic odor when struck;distinguished from smaltite by blowpipe tests.	Widespread and associated with the common sulfides.	Pyrite Marcasite Smaltite	Principal source of white arsenic.
2½-5½; varied.	Colorless	Separation into fine, slender fibers that are somewhat flexible.	Alteration product from serpentines & amphiboles.	Zeolites	Manufacture of asbestos products.
3½-4	Light-blue.	Azure-blue color; crystalline occurrence;small water content;softer than its similarities.	Oxidized zone of copper deposits. commonly associated with malachite.	Sodalite Lazurite Lazulite	Ore of copper; gem material.
2½-3½	Colorless	Breaks into tabular forms; high sp.gr.; does not effervesce in acids.	In separate veins and as a gangue mineral with deposits of galena. sphalerite, etc.	Dolomite Calcite Celestite Strontianite	Paint pigments; in barium salts;& as a heavy medium in well muds.
1-2 Soft; red types harder	Non-metallic	Clayey odor and fracture; chemical analyses are necessary for exact identification.	A weathering product, probably from the desilication of clay.	Clay Laterite	Ore of aluminum; refractories.

Name	Color	Luster	Cleavage	Fracture	Form	Sp. Gr.
BENTONITE Complex silicate	Light colors; varied	Earthy	None	Earthy & conchoidal like clay	Similar to that of clay	2.5
BERYL BeAl silicate	Emerald green; blue & yellow	Vitreous	Imperfect and indistinct	Conchoidal to uneven; brittle	Hexagonal crystals with a flat base; rarely as columnar masses.	2.7
BIOTITE Complex silicate	Dark brown & greenish black	Shiny on cleavage surfaces	Perfect in 1 direction	Uneven; brittle.	Thin sheets; massive aggregates of cleavable scales or flakes.	2.9
BORAX $Na_2B_4O_7 \cdot 10H_2O$	White when pure; grey	Vitreous to dull	Imperfect	Uneven	Individual crystals; aggregates of crystals; and glassy masses.	1.7
BORNITE Cu_5FeS_4	Copper red with purple tarnish	Metallic	None	Uneven; conchoidal; brittle.	Compact masses and scattered specks	5.
CALAMINE $Zn_2(OH)_2 SiO_3$	Colorless; white; pale colors	Vitreous to pearly	1 direction, perfect	Uneven to sub-conchoidal; brittle.	Massive; colloform; rounded; tabular crystals.	3.4 to 3.5
CALCITE $CaCO_3$	Colorless to any light color	Vitreous	Perfect in 3 directions, forming rhombs	Indistinct; brittle	Separate crystals of varied habits; crystalline masses	2.7
CASSITERITE SnO_2	Brown or black; rarely varied	Submetallic to adamantine	Imperfect	Sub-conchoidal; brittle.	Crystals; stream-worn pebbles; & in mammillary nodules.	7.
CERARGYRITE AgCl	White; grey; green	Resinous to adamantine	None	Unimportant; sub-conchoidal	Massive; thin crusts or coatings	5.5
CERUSSITE $PbCO_3$	Grey; brown; yellow	Adamantine to dull	Not prominent	Conchoidal; brittle.	Compact masses; fibrous; rarely in crystals	6.5
CHALCEDONY SiO_2	White to varied	Waxy & brilliant to dull	None	Conchoidal; brittle	Massive and banded concretionary nodules.	2.6

Hard-ness	Streak	Distinguishing Features	Occurrence	Similar-ities	Uses
1-2; soft.	White	Dry surfaces adhere to the tongue; clayey odor most varieties swell when placed in water.	A bedded clay deposit that is usually derived from the weathering of volcanic ash.	Other clay minerals & bauxite.	Filter; coagulant; molding ingredient in sands.
7½-8	None	Light colors;harder than apatite;softer than corundum;some crystals show vertical striations.	In granitic pegmatites, mica schists,& gneisses.	Quartz Apatite Corundum Chrysoberyl	Source of beryllium metal, gems
2½-3	White to light green	Easy cleavage dark even in thin flakes; flexible.	In igneous rocks,granites, pegmatites, sands,& metamorphic rocks.	Phlogo ite Vermiculite Chlorite	Flake or powder used in asphalt roofing.
2-2½	White	Soluble in water; feeble sweetish alkaline taste	Salt deposits of evaporated lakes, with other borates.	Borates Epsomite Alum	Source of boron; washing compound..
3	Grayish-black	Distinct purple tarnish; softer than other iron copper minerals.	Widespread in vein deposits & igneous magmas associated with other sulfides.	Chalcocite Covellite Chalco-pyrite	An important ore of copper.
4½-5	White	Cleavage, which smithsonite does not have.	Oxidized zone, with smithsonite; derived from sphalerite.	Smithsonite	Ore of zinc.
3	White	Cleavage rhombs; bubbling in cold, dilute muriatic acid (HCl).	Widespread as secondary mineral;primary in sedimentary deposits.	Aragonite Barite Dolomite Siderite	Optics; same uses as lime-stone.
6-7	Grey; rarely brown	High sp.gr.;hard ness;infusibility; streak.	Primary in granitic pegmatites;placer deposits.	Rutile Wolframite Tourmaline Garnets	Ore of tin
1-1½; highly sectile	White	Sectility; waxy and greasy appearance	Secondary mineral found in upper parts of silver deposits	Wavellite and some copper carbonates	Ore of silver
3-3½	Uncolored	High sp.gr.;luster; bubbling in dilute nitric acid.	Secondary mineral in oxidized zones of lead deposits.	Anglesite Barite Witherite	Minor ore of lead
7	None	Hardness; conchoidal fracture; and fine grain	Cavity and vein fillings common in lava flows.	Quartz Chert Opal	Gem stone; abrasive

Name	Color	Luster	Cleavage	Fracture	Form	Sp. Gr.
CHALCOCITE Cu_2S	Sooty-black and dull grey	Metallic	Indistinct	Granular; imperfect; conchoidal	Compact masses; sometimes granular.	5.7
CHALCOPYRITE $CuFeS_2$	Brass-yellow	Metallic	None	Sub-conchoidal; brittle	Compact masses; separate crystals that look like tetrahedrons	4.2
CHERT SiO_2	White; grey; red & black	Dull	None	Conchoidal; brittle	Rounded nodules; lenses; & stratified beds	2.6
CHLORITE Complex silicate	Dark-green	Vitreous to pearly	One direction; basal perfect but not as good as mica.	Splintery	Foliated and scaly masses.	2.6-2.96
CHROMITE $FeO.Cr_2O_3$	Brownish black	Metallic	None	Uneven	Commonly in compact masses and granular aggregates.	4.4
CHRYSOBERYL	Distinguished from beryl only by chemical and optical					
CHRYSOCOLLA $CuSiO_3 2H_2O$	Pale-blue to greenish-blue	Vitreous to dull	None	Conchoidal to earthy; brittle	Cryptocrystalline; enamel-like in texture	2.2
CINNABAR HgS	Orange; crimson; & brownish red	Adamantine to dull.	Rhombohedral Perfect in 3 directions; unimportant	Uneven	Small crystals; crystal aggregates; massive & earthy forms	8
CLAY Hydrous aluminum silicate	Varied, usually light colored	Dull to earthy	None	Uneven; conchoidal in flint clays	Massive & bedded	2.6
COAL Solid hydrocarbon	Black to brown	Dull to vitreous	None	Even and conchoidal	Jointed blocks & angular fragments	1.2-1.7
COBALTITE $CoAsS$	White to greyish-black, reddish tinge	Metallic	3 directions, perfect, at right angles	Uneven	Massive; crystals like pyrite.	6 to 6.3
COVELLITE CuS	Indigo-blue or darker	Sub-metallic to resinous	Basal, perfect.	None	Usually thin hexagonal plates; massive.	4.6

Hard-ness	Streak	Distinguishing Features	Occurrence	Similar-ities	Uses
2½-3; sub-sectile	Black	Sub-sectility;soft-ness; and sooty appearance	Enriched sul-fide zones of copper veins; with other cop-per sulfides.	Argentite Bornite Tetra-hedrite	High grade ore of copper
3½-4	Greenish black	Softer than pyrite; non-magnetic;purple and copper sheen.	Common as a primary mineral with other Cu-Fe sulfides.	Pyrite Pyrrhotite Bornite Gold	Most important ore of copper.
7	None	Hardness; typical conchoidal fracture; & very fine grained.	In sedimentary beds with chalk, limestone, and shale.	Quartz Chalcedony	None
1-2½	Greyish green	Scaly flakes with fair elasticity; green color;and greasy feel.	Alteration product of ferro-magne-sium minerals	Mica Talc	None
5½	Greyish brown	Feebly magnetic; dark brown streak; association with serpentine and magnetite.	Basic igneous & metamorphic rocks,such as peridotite and greenstone.	Franklinite Magnetite Ilmenite Specular-hematite	Ore of chromium

tests.

2-4	White to light blue	Light-blue color; glassy; inferior hardness;and non-effervescence.	Secondary min-eral in leached and oxidized zones.	Turquoise Malachite Azurite	Minor ore of copper; gem stone.
2-2½	Scarlet or a dark red.	Bright red color; scarlet streak;and high sp.gr.	Primary mineral deposited as veins & specks in various kinds of rocks.	Hematite Cuprite Realgar	Important ore of quick-silver.
1½-2½; soft	Uncommon; red var-ieties are yellowish	Clayey odor; dry clay sticks to tongue;sticky and soapy when wet.	Alteration pro-duct from dif-ferent rocks; usually resid-ual deposits.	Bauxite Shale Siltstone Bentonite	Ceramic industry in making pot-tery,bricks, etc.
1-1½	Brown to black.	Black color & streak; brittleness;& com-bustibility.	Interbedded in sed.rocks and in pockets of sed.rocks.	Oil shale and other solid hy-drocarbons	Fuel.
5½	Greyish-black	Cubic cleavage; white color as op-posed to pyrite's yellow.	In veins with other cobalt & nickel miner-als.	Arsenopy-rite Smaltite Pyrite	Ore of cobalt.
1.5-2	Lead-grey to black	Often shows fine purple color when moistened with water.	Associated with other copper minerals;usual-ly secondary.	None	Ore of copper.

Name	Color	Luster	Cleavage	Fracture	Form	Sp. Gr.
CUPRITE Cu_2O	Dark red & brownish red	Adamantine to submetallic.	Imperfect	Uneven	Crystals; crystalline aggregates; and fine-grained masses.	6.0
DIAMOND C	Colorless to black	Adamantine to greasy	Perfect in 4 directions forming octahedrons	Conchoidal	Loose, single, & twinned crystals, with curved faces and edges.	3.5
DIATOMITE Hydrous silica	White & grey, colored by impurities	Dull to earthy	None	Uneven	Massive bedded deposits with a powdery, chalk-like appearance.	1.9-2.35
DOLOMITE $CaMg(CO_3)_2$	White & light grey	Pearly to Vitreous	3 directions at oblique angles; rhomb faces curved.	Unimportant	Crystal coatings & aggregates; vein fillings; massive in beds.	2.8-3.0
EPIDOTE Complex Ca, Al, & Fe silicate	Pistachio green; yellowish green.	Vitreous	1 direction, basal perfect.	Uneven; brittle.	Crystals, columnar aggregates; and granular masses.	3.2-3.5
FELDSPAR (K Na, or Ca) Al silicate	White to grey to pink; green; & blue.	Vitreous & pearly	2 directions almost at right angles, sometimes only 1 cleavage distinct.	Uneven across cleavage planes	Crystals, loose & embedded in rock masses intergrown cleavable aggregates.	2.5-2.7
FLUORITE CaF_2	Transparent; light colored tints.	Vitreous	4 directions, forming perfect octahedrons.	Sub-conchoidal; compact kinds are splintery; brittle.	Cubic crystals; sometimes octohedrons; granular masses.	3.0-3.2
GALENA PbS	Lead-grey; greyish black.	Metallic	Perfect in 3 directions; small cubes break from large cubes.	Undeveloped.	Common in cubes; rare in octahedrons; in ore bodies as cleavable masses.	7.5; Heavy
GARNET Complex silicate	Varied; red to black.	Vitreous to resinous	Usually absent but fracture parting is common.	Uneven to sub-conchoidal	Distinct 12-sided crystals; granular masses; rounded grains in sand.	3.1-4.3
GOLD Au	Pale yellow; silvery tinge.	Metallic	None	Hackly; placer gold rounded; malleable & ductile.	Finely disseminated thru the rock; flakes; wire stringers; rounded nuggets	15-19

Hardness	Streak	Distinguishing Features	Occurrence	Similarities	Uses
$3\frac{1}{2}$-4	Brick red and cochineal.	Absence of perfect cleavage;bright red streak;& hardness	Oxidized zone of copper deposits;with native copper.	Cinnabar Hematite	Minor ore of copper.
10	None	Extreme hardness; brilliant light dispersion;greasy appearance of rough stone.	Stream gravels; and kimberlite (altered peridotite).	Quartz Topaz Zircon Corundum	Precious gem; abrasives; and in machine tools.
1-$1\frac{1}{2}$; soft	None	Light weight;soft chalky & clayey feel when rubbed or crushed between the fingers.	Bedded basin-shaped deposits associated with sediments and volcanic lavas.	Pumicite Chalk Tripoli Clay	Filter aid; insulant; admixture in concrete.
$3\frac{1}{2}$-4	None	Curved cleavage faces;weak effervescence in cold dilute hydrochloric acid.	Similar to calcite (limestone) in sed. beds;vein and cavity fillings.	Calcite Magnesite Siderite Rhodochrosite	Building & ornamental stone; refractory.
6-7	Grey to pale yellow	Distinct pistachio green color;deep striations;reticulated masses.	All kinds of high lime-bearing metamorphic rocks.	Amphiboles Pyroxenes Tourmaline Feldspar	None
$5\frac{1}{2}$-$6\frac{1}{2}$	White or absent.	Crystalline structure;good cleavage; twin intergrowths; twinning lines or striations on the plagioclase felispars.	All classes of rocks;abundant in acid igneous rocks,granites & pegmatites.	Apatite Epidote Rhodonite	Ceramics; scouring soaps; glass industry.
4	White	Good octohedral cleavage; harder than barite,celestite,& calcite; softer than apatite.	Gangue mineral in lead and zinc sulfide deposits;vein mineral in limestone.	Calcite Barite Celestite Strontianite Apatite	Flux in electric furnaces; glass & chemical industries.
$2\frac{1}{2}$	Lead-grey	High sp.gr.;perfect cubic cleavage;lead-grey color.	In vein,contact metamorphic & replacement deposits,with other sulfides.	Stibnite Argentite	Most important ore of lead.
6-$7\frac{1}{2}$; crystals tough	White	Hardness;absence of cleavage;12-sided crystal habit.	Common accessory mineral in metamorphic rocks,limestones and sands.	Zircon Spinel Epidote Corundum Staurolite	Important as an abrasive; gem stone.
$2\frac{1}{2}$-3	Pale yellow; similar to color.	Uniform non-flickering yellow color; high sp.gr.;malleability & ductility.	Widely distributed in quartz veins; in stream & beach placers.	Pyrite Chalcopyrite Golden mica	Monetary; in the arts.

Name	Color	Luster	Cleavage	Fracture	Form	Sp. Gr.
GRAPHITE C	Steel-grey to black.	Metallic to dull.	Basal, which is perfect in 1 direction.	Unimportant	Foliated or granular masses; minute disseminated scales.	2.2
GYPSUM $CaSO_4 \cdot 2H_2O$	White & grey; tinted shades.	Pearly & silky; sometimes dull.	Perfect in 1 direction; imperfect and fibrous in the others.	Fibrous & sub-conchoidal.	Single platy crystals; foliated, fibrous, columnar, & granular masses.	2.3
HALITE $NaCl$	Transparent or white; tinted.	Vitreous	Cubic perfect, and easily developed.	Conchoidal; brittle.	Cubic crystals & cubic fragments; granular & compact masses.	2.1-2.6
HEMATITE Fe_2O_3	Steel-grey; red; & black.	Metallic	None	Hackly; sometimes conchoidal.	Massive; mammillary; fibrous; columnar; oolitic; and foliated masses.	5.0
ILMENITE $FeTiO_3$	Iron-black	Metallic	None	Conchoidal brittle	Tabular crystals; flat plates; compact and granular masses; sand.	4.5-5.0
LIMONITE $2Fe_2O_3 \cdot 3H_2O$	Brown to yellowish brown.	Dull to submetallic.	None	Uneven; conchoidal.	Compact; botryoidal; nodular; and pisolitic masses.	3.8
MAGNESITE $MgCO_3$	Snow-white to grey	Vitreous to dull.	Unimportant; rhombic in crystalline varieties.	Conchoidal in fine-grained varieties.	Common in compact porcelain-like masses; sometimes coarsely crystalline and granular.	3.0
MAGNETITE Fe_3O_4	Black	Metallic; sometimes dull.	Not distinct; occasional octahedral parting.	Sub-conchoidal; brittle.	Small crystalline or non-crystalline grains; granular masses; sand.	5.1
MALACHITE $Cu_2(OH)_2 CO_3$	Emerald green	Vitreous to dull; velvety.	Unimportant.	Uneven to sub-conchoidal; brittle.	Massive or incrustations with botryoidal or fibrous or stalactitic surfaces; acicular crystals.	3.9-4
MANGANITE $Mn_2O_3 \cdot H_2O$	Black or dark grey.	Submetallic.	1 or 2 directions parallel to length of crystal.	Uneven	Prismatic crystals vertically striated; columnar and fibrous masses.	4.3
MICA	(see Biotite and Muscovite)					

Hard-ness	Streak	Distinguishing Features	Occurrence	Similar-ities	Uses
1-2 soft greasy feel	Black or dark, steel-grey	Softness;greasy feel;black streak; rubbing forms black marks on the fingers.	Widespread in nearly all kinds of metamorphic rocks.	Molybdenite. Specularite.	Mfg.of pen-cils, crucibles, lubricants.
1½-2	White	Softness;1 perfect cleavage and the others imperfect or fibrous;non-effer-vescent in acid.	Bedded deposits associated with sedimentary rocks.	Anhydrite Calcite ZEolites Mica Talc	Calcined gypsum; soil stabilizer; flux.
2½	White	Perfect cubic cleav-age;softness; and saline or salty taste.	As solid salt in beds & salt domes;in briny solutions.	Sylvite Carnallite Fluorite	Table salt; chemical & glass industry.
5½-6½	Brownish red or dark red.	Brownish-red streak; high sp.gr.;contains no water or very little water.	Widespread and associated with basic igneous rocks & meta-morphic rocks.	Limonite Magnetite Chromite Ilmenite Franklinite	Most im-portant ore of iron.
5-6	Brownish-black.	Brownish-black streak;weakly magne-tic or non-magnetic.	In basic coarse-grained igneous rocks and in black sands.	Magnetite Hematite Chromite Wolframite	Ore of titanium Paint filler
1-5½	Yellowish brown.	Yellowish-brown streak;high water content;compact structure.	Alteration pro-duct of various iron minerals; widespread.	Magnetite Hematite Chromite	Minor ore of iron; yellow pigment.
3½-4½; sometimes harder.	White	Fairly high sp.gr. & hardness;tough-ness;fine-grained texture;mild effer-vescence.	As lenses in sediments & as small deposits in metamorphic rocks.	Calcite Dolomite Rhodochro-site.	Caustic and dead-burned magnesite; refractory.
5½-6½	Black	Strong magnetism; black streak;bril-liant black color.	Widely distrib-uted in basic igneous rocks & metamorphic rocks.	Ilmenite Franklinite Hematite Chromite	Ore of iron
3½-4	Pale-green	Uniform green color; fibrous or acicular crystal structure; effervescence in HCl.	Characteristic in oxidized zone of copper dep-osits;wide-spread.	Chrysocolla Turquoise Smithsonite	Minor ore of copper; gem stone.
4	Reddish brown to black.	Crystalline struc-ture;softer than psilomelane;harder than pyrolusite.	Secondary min-eral in veins & cavities; residual in clays.	Psilomelane Pyrolusite Limonite Hematite	Ore of manganese.

Name	Color	Luster	Cleavage	Fracture	Form	Sp. Gr.
MOLYBDEN-ITE MoS	Bluish lead grey	Metallic bright	Almost mica-ceous	None	In foliated masses	4.7 to 4.8
MONAZITE (Ce,La,Nd, Pr) PO$_4$	Yellow; brown; reddish	Resin-ous	Parting in 1 direction	Conchoid-al to uneven.	Grains in various sands.	4.7 to 5.3
MUSCOVITE complex silicate.	White, green & yellow	Vitreous to pearly	Perfect in 1 direction, splitting into thin sheets	Hackly on cleavage edges.	Foliated flakes & scaly masses; large sheets and "books"; tabular structure.	2.76 to 3.0
NICCO-LITE NiAs	Pale copper-red	Metallic	None	Uneven	Nearly always in massive form.	7.3 to 7.7
NITRATINE (Nitrates)	White or color-less.	Earthy	Rhombohedral (3 directions at oblique angles).	Not ap-parent.	Rhombohedral crys-tals, usually massive.	1.5 to 2.
OLIVINE (Mg,Fe)$_2$ SiO$_4$	Yellow-green	Vitreous	Indistinct	Conchoid-al;brittle	Granular masses.	3.3
OPAL SiO$_2$-H$_2$O	Varied	Greasy to pearly	None	Conchoid-al.	Nodular masses.	2. to 2.3
ORPIMENT As$_2$S$_3$	Lemon-yellow	Dull	Not distinct	None	Foliated masses or powdery incrustations.	3.4 to 3.5
ORTHOCLASE (K,Na)AlSi$_3$ O$_8$	White, green, pink	Vitreous to pearly	2 directions at right angles with parting at an oblique angle.	Uneven	In cleavable crys-tals and masses embedded in rock.	2.5 to 2.6
PITCHBLENDE	Grey-green to black	Pitchy or greasy to dull	None	Conchoid-al to un-even	Massive	6.5 to 8.0
PLAGIO-CLASE (Na,Ca) AlSi$_3$O$_8$	White to grey	Vitreous	1 direction perfect, 1 direction less perfect.	Uneven	In cleavable crys-tals and masses embedded in rock.	2.6 to 2.7
PLATINUM Pt	Silvery white	Bright metallic	None	Hackly	Rounded grains and nuggets; irregular lumps	14 to 19

Hardness	Streak	Distinguishing Features	Occurrence	Similarities	Uses
1-1½	Green to blue-black; bluish grey on paper	Streak is more greenish than graphite and luster brighter	In pegmatites, contact metamorphic zones. High temp. quartz veins.	Graphite	Ore of molybdenum.
5	None	Cleavage, if grains are large enough to show cleavage.	In certain sand deposits derived from salic igneous rocks.	Titanite Zircon	In some cases, a source of thorium.
2-2½	Colorless	Highly flexible & elastic;splits into very thin sheets;	Pegmatites; most light colored igneous rocks;metamorphic rocks.	Gold flakes Chlorite Gypsum	Electrical insulation non-conductor.
5-5½	Pale brownish-black.	Its copper sheen; otherwise chemical tests are necessary.	Occurs with nickel & cobalt minerals.	Cobaltite Pyrite	Ore of nickel
1½-2	White	Cool, saline taste.	Arid regions by evaporation of water.	None	Fertilizer; manufacture of potassium nitrate
6½-7	Colorless	Greenish color and absence of cleavage	Basic or femic igneous rocks; such as peridotite & basalt.	No common ones	Refractory; gem
5½-6½	Colorless	Softness and low specific gravity; greasy luster	Deposited from siliceous solutions as secondary minerals.	Chalcedony	Certain varieties as gem material
1½-2	Pale lemon-yellow	Yellow color.	Vein mineral often associated with realgar	None	Synthetic orpiment as a paint pigment
6-6½	Non-metallic	Cleavage, carlsbad twinning	In salic igneous rocks & pegmatites;as grains in some sandstones	Plagioclase	Ceramics
5½	Olive-green to black	High specific gravity	Metalliferous veins with metallic-sulfide minerals.	Obsidian Pitchstone	Source of uranium and radium.
6	Colorless	Cleavage and polysynthetic twinning.	In more basic (femic)rocks and as grains in sandstone.	Orthoclase	Ceramics
4½; malleable.	Light steel grey	High specific gravity; malleability	Assoc.with peridotite or in placers derived from them.	Similar to bright iron or steel shavings	Jewelry; scientific uses; catalyst.

Name	Color	Luster	Cleavage	Fracture	Form	Sp. Gr.
PROUSTITE Ag_3AsS_3	Scarlet-vermilion	Adamantine	1 direction good	Conchoidal to uneven	Massive crystals	5.57 to 5.64
PSILOME-LANE $4MnO_2$ (Ba,K_2) $O(H_2O)_x$	Black to brown	Sub-metallic to dull	None	Conchoidal brittle.	Compact earthy material that is amorphous.	3.7 to 4.7
PYRARGY-RITE Ag_3SbS_3	Black to dark red	Metallic; adaman-tine	1 direction good	Conchoidal to uneven	Massive; prismatic crystals	5.7 to 5.8
PYRITE FeS_2	Silvery to pale brass yellow	Metallic	Imperfect cubic	Uneven to conchoidal	Massive and granular forms; in cubes and pyritohedrons	5
PYRO-LUSITE $MnO_2(H_2O)_x$	Black	Metallic to dull	None	Splintery, not often distin-guishable	Fibrous and colum-nar forms; in crusts and masses.	4.8
PYROXENE Complex silicate	Black to green to white; usually darker colors	Vitreous to pearly.	Imperfect in 2 directions at nearly right angles; parting in 1 direction at an oblique angle.	Uneven; brittle	Embedded crystals, usually stubby; bladed and colum-nar masses.	3.2 to 3.6
QUARTZ SiO_2	White to varied	Vitreous	Unimportant	Conchoidal	Massive (bull quartz; hex.crystals with pointed ends.	2.6 to 2.7
REALGAR As_2S_2	Red to orange-red.	Adaman-tine to resinous.	Not promin-ent	Conchoidal	Massive; granular; crystals.	3.5
RHODOCHRO-SITE $MnCO_3$	Pink to brownish red	Vitreous to pearly	3 direc-tions at oblique angles like cal-cite	Uneven; brittle	Crystals and fine to coarse grained cleavable masses.	3.5 to 3.6
RHODONITE $MnSiO_3$	Pink to red	Vitreous	2 directions at nearly right angles.	Conchoidal to uneven	Cleavable and com-pact masses	3.4 to 3.6

Hardness	Streak	Distinguishing Features	Occurrence	Similarties	Uses
2-2½	Scarlet-vermil-ion	Lighter color and streak than pyr-argyrite.	Vein mineral formed by as-cending sol-utions	Pyrargyrite Cuprite Cinnabar	Ore of silver
5-6; tough.	Brownish black to brown.	Absence of crys-talline struc-ture & streak.	In residual clays and bog deposits.	Pyrolusite Manganite Limonite	Ore of man-ganese.
2½	Purplish red	Darker color and streak than prous-tite.	Vein mineral formed by as-cending sol-utions.	Proustite Cuprite Cinnabar	Ore of silver
6-6½	Greenish to brownish black	Harder than pyr-rhotite and chalco-pyrite;a low temp. form called mar-cassite is very similar.	Assoc.with all sulfides & in all rocks; of-ten in tiny cub-ic crystals	Pyrrhotite Chalcopyrite Gold Marcasite Arsenopyrite	Mfg.sulfuric acid
2-2½	Black	Inferior hardness and low moisture content.	Similar to other manga-nese oxides,pro-bably derived from mangan-ite.	Psilomelane Manganite Limonite	Ore of man-ganese
5-6	Color-less	Contrasted with amphibole-cleavage at 90o instead of 60°; square in-stead of hexagonal cross section;stub-by instead of elon-gated.	In femic (basic) igneous rocks; do not common-ly occur in rocks containing quartz.	Amphibole Epidote Tourmaline	None important.
7	None	Its high hardness and lack of cleav-age.	Salic igneous rocks;veins; grains in sand-stone & some metamorphics.	Calcite Scheelite Feldspars	Abrasive; glass making; gems; flux.
1½-2; sectile	Orange-yellow	Lower specific gravity and hard-ness than cinnabar.	Vein mineral associated with stibnite,cinna-bar,pyrite,etc.	Cinnabar	Source of arsenic.
3½-4½	Color-less	Heavier than cal-cite and dolomite and softer than rhodonite.	Vein mineral with silver ores and with some Mn ores.	Calcite Dolomite Magnesite Rhodonite	Ore of man-ganese
5½-6½; tough	Color-less	Heavier than ortho-clase & harder than rhodochrosite;cleav-age differs from pink carbonates.	In high temp. veins, and in crystalline limestones	Rhodochro-site Calcite etc.	Ornamental stone and gem material.

Name	Color	Luster	Cleavage	Fracture	Form	Sp. Gr.
RUTILE TiO_2	Brownish red to black.	Adamantine; metallic	Not important.	Uneven; brittle	Embedded grains or crystals	4.2
SCHEELITE $CaWO_4$	White, grey and pale	Sub-adamantine	Distinct in 1 direction with step-like interruptions.	Uneven; brittle	As crystals and in massive form.	6
SERPENTINE Complex silicate	Greenish, brownish, reddish, variegated	Resinous to greasy	Not important.	Conchoidal to splintery	Massive	2.5 to 2.65
SIDERITE $FeCO_3$	Various shades brown & grey	Sub-adamantine to high vitreous	3 directions at oblique angles; (rhombohedral) like calcite	Uneven; brittle.	Small crystals; cleavable masses; crusts; concretions; and oolites	3.8
SILVER Ag	Tin white to pale yellow	Metallic	None	None	Wire; thin sheets; skeleton crystals; rarely as nuggets.	10.1 to 11.1
SMALTITE (Co,Ni) As_2	White to grey; iridescent	Metallic	None	Granular	Massive; cubic crystals	5.5 to 6.8
SMITHSONITE $ZnCO_3$	White to varied; sometimes blue or green.	Vitreous	3 directions at oblique angles (rhombs), but not prominent.	Uneven; brittle.	Nodular masses and crusts.	4.3 to 4.5
SPHALERITE ZnS	Pale yellow to black, dependent on iron content.	Resinous to vitreous	Dodecahedral, 12 sided figure derived from cube.	Conchoidal brittle	Massive; crystals	4.0
STIBNITE Sb_2S_3	Dark lead-grey	Metallic to dull on fresh surface	Perfect in 1 direction with cross lines on cleavage surfaces.	Uneven	Granular masses; slender crystals in radiating groups.	4.5

Hardness	Streak	Distinguishing Features	Occurrence	Similarities	Uses
6-6½	Pale brown	Lower sp.gr.than cassiterite; red color and adamantine luster.	High temp. veins or pegmatites; in metamorphics.	Cassiterite	Coloring matter for porcelain & production of ferrotitanium.
4½-5	White	High sp.gr.and sub-adamantine luster with disseminated masses; ultra-violet lamp is helpful.	In veins formed at high temperature.	Quartz	Ore of tungsten.
2½-4	White, slightly shining	Softness,greasy luster & "feel"; absence of cleavage.	Alteration product of femic rocks and of amphibolites.	Talc Chlorite	Ornamental stone
3½-4	Pale brown or grey	Higher sp.gr. than calcite; rhomb.cleavage and high luster from sphalerite	Vein mineral; replacement of limestone concretions; small crystals in cavities.	Calcite Dolomite Sphalerite	Minor ore of iron
2½-3 Malleable	Metallic, silver-white	Bright color of freshly cut surface and malleability	Vein min.with cobalt & nickel & other silver minerals;in the "iron hat" or gossan,rarely in placers	Resembles some fresh surfaces of native metals as lead & tin.	Silver; jewelry; monetary uses.
5½-6	Greyish-black	Lack of cleavage; otherwise chemical tests are necessary.	With silver and copper ores;in veins with ores of cobalt,nickel & copper.	Arsenopyrite Cobaltite Pyrite	Ore of cobalt and nickel.
5	Colorless	Harder than other carbonates; harder and heavier than chrysocolla.	In oxidized zone and formed from alteration of sphalerite.	Other carbonates. Chrysocolla	Minor ore of zinc.
3½-4	Pale yellow to dark brown	Perfect cleavage and high luster; softer than garnet.	Vein mineral with other sulfides; replacement mineral in sed.rocks; a contact metamorphic mineral.	Garnet Siderite	Ore of zinc.
2	Dark lead grey to black.	Perfect cleavage; softness; low fusibility.	Associated with other sulfides.	Galena	Principal source of antimony.

Name	Color	Luster	Cleavage	Fracture	Form	Sp. Gr.
SULFUR S	Yellow to yellowish red.	Resinous to adamantine.	Absent	Brittle	Crystals; crusts; masses.	2.
TALC Complex silicate	White to grey to pale green	Pearly	Perfect in 1 direction into foliated masses.	None	Scales; foliated; fibrous; and compact masses.	2.7
TETRAHEDRITE Copper-antimony sulfide	Dark iron grey	Metallic	Absent	Uneven, brittle	Usually massive; tetrahedral crystals	4.4 to 5.1
TITANITE or SPHENE CaTiSiO5	Varying tints of yellow or brown.	Adamantine to high vitreous	Indistinct	None; brittle	Flat, wedge shaped crystals; massive; granular.	3.4 to 3.5
TOURMALINE Complex silicate.	Black; brownish; bluish; green; red.	Vitreous to resinous	None	Sub-conchoidal to uneven.	Prismatic crystals	3 to 3.25
WOLFRAMITE (Fe,Mn) WO4	Black or dark brown.	Metallic to sub-metallic	One direction perfect.	Uneven; brittle.	Heavy black masses; tabular crystals.	7.2 to 7 5
ZIRCON ZrSiO4	Reddish-brown to light variegated.	Adamantine	Imperfect	Conchoidal	Usually found as loose grains in some "black sands"	4.7

Hardness	Streak	Distinguishing Features	Occurrence	Similarities	Uses
$1\frac{1}{2}$-$2\frac{1}{2}$	Colorless	Yellow color; low hardness; brittleness; odor of sulfur when burned.	Decomposition of many sulfides; product of vulcanism.	None	Manufacture of sulfur products.
1-$1\frac{1}{2}$; may range to 4 with impurities	White Nonmetallic	Soapy feeling & low hardness.	Results from alteration of amphiboles, and serpentine.	Mica Chlorite	Talcum powder; paper glaze; refractory.
3-$4\frac{1}{2}$	Dark grey	Chalcocite is sectile while tetrahedrite is brittle.	Vein mineral with other sulfides.	Chalcocite	Ore of copper
5-$5\frac{1}{2}$	White	Crystal shape.	Common in granitic rocks with magnetite & hornblende; in metamorphic rocks.	Magnetite Hornblende	None important.
7-$7\frac{1}{2}$	Colorless	Triangular crystal cross-section and lack of cleavage.	In salic igneous rocks and in certain metamorphic rocks.	Amphibole Pyroxene Epidote	Transparent varieties as gems.
5-$5\frac{1}{2}$	Black to dark reddish brown.	High sp.gr. and cleavage in only 1 direction.	In granular salic rocks & pegmatites; assoc. with cassiterite & scheelite.	Hubnerite	Ore of tungsten
$7\frac{1}{2}$	Colorless	Difficult without optical tests.	Loose crystals in sands derived from salic rocks.	Diamond	Refractory; ferro-alloys; gems.

GLOSSARY

The definitions given herein are taken principally from "A Glossary of the Mining and Mineral Industry". by Albert H. Fay, published by the U.S. Bureau of Mines as Bulletin 95, and from Webster's New International Dictionary, 2d Edition.

Aberration: A lens condition that permits light rays to be focussed at different points not all on the same plane or field of observation. It is impossible to bring the entire "field" in focus if a lens is not properly corrected.

Spherical Aberration: is the inability of the marginal rays of a lens to focus in the same plane with those passing through the center of the lens. As a result, the center of the "field" may be in focus but the margins are out of focus.

Chromatic Aberration: is caused by the inability of the lens to focus different colors of light in the same plane. The "field" is surrounded by rings of color.

Acicular: needle shaped; slender, like a needle or bristle, as some leaves or crystals.

Acidic Rocks: a descriptive term applied to those igneous rocks that contain more than 65 percent silica (SiO_2) as contrasted with intermediate or basic rocks. The term "Salic rocks" is preferable to "Acidic rocks".

Adamantine: a term used to describe the luster of a mineral; luster like that of a diamond.

Adsorb: to condense and hold a gas on the surface of a solid, particularly metals. Also to hold a mineral particle within a liquid interface. During the World War (1914-1918) carbon made from peach stones was capable of holding gas to the surface of the carbon particles. No gas was absorbed, however.

Alumina: oxide of aluminum, Al_2O_3.

Amorphous; a term applied to rocks and minerals having no definite molecular structure.

Amygdules: a small globular cavity in an eruptive rock such as lava, caused by expanding steam or vapor at the time the rock cooled; they generally are lined or filled with secondary minerals.

Argentiferous: containing silver.

Basalt: the basalt or basaltic group of rocks is used to include all the dark, femic, fine-grained volcanic rocks, such as the true basalts and others.

Basic: this term as applied to rocks is a general one for those igneous rocks that are comparatively low in silica. About 45 or 50 percent of silica is the upper limit. The term "femic rocks" is preferred to "basic rocks".

Bladed: decidedly elongated and flattened, like a knife blade.   The term is
    used to describe the structure of certain minerals.

"Books": a term used to describe mica crystals from which individual sheets can
    be stripped.   The similarity of these mica sheets to the leaves of a book
    explains the term.

Botryoidal: having the form of a bunch of grapes.   A mineral has botryoidal form
    if the surface is a series of bumps, each one about the size of a grape.

Carlsbad Twinning: a type of twinned crystal occurring in the monoclinic crystal
    system with the vertical axis as the twinning axis.

"Colors":   refers to particles of metallic gold.

Conchoidal: shell shaped.   The more compact rocks such as flint, argillite, fel-
    site, etc., break with conchoidal fracture.   The curved, or conch-like breaks
    on the surface of a broken chunk of glass are examples of conchoidal fracture.

Contact Metamorphism: a general term applied to the change that takes place along
    an intrusive contact (of an intruded igneous rock and the enclosing rocks into
    which it has been thrust), such as the recrystallization of limestone with the
    development of typical silicate minerals.

Country Rock: the general mass of adjacent rock as distinguished from that of a
    dike, vein, or lode.

Cryptocrystalline: formed of crystals of microscopic fineness, but not glassy.

Diameters: a unit of magnification of microscopic and telescopic observation
    equal to the number of times the linear dimension of an object is increased;
    as, a magnification of eight diameters means the dimensions are increased
    in the ratio of 8 to 1.

Disseminated: to be scattered or diffused through.   A disseminated mineral is
    one which is scattered through the rock rather than being concentrated in
    masses or veins.

Dodecahedron: an isometric (cubic) form composed of 12 faces, each parallel to
    one axis and intersecting the other two axes at equal distances.

Effervescent: to bubble and hiss, as limestone or any carbonate on which acid
    is poured.

Femic: a coined word, portions of ferromagnesium, that describes those rocks or
    minerals that have a predominance of ferromagnesian minerals.   Synonomous
    with "basic" when basic is applied to rocks.

Field Magnification: a sphere or range of activity or observation,...the area,
    usually circular, visible through the lens system of an optical instrument,
    as a microscope or telescope.

Fluorescence: the emission of light from within a substance while it is being ex-
    posed to direction radiation, or in certain cases, to an electrical discharge,

(as from an ultra-violet lamp). A fluorescent substance gives off light when excited with ultra-violet rays.

Flux: material that will assist fusion by forming more fusible compounds.

Foliated: leaf-like. The meaning is similar to that of laminated, but the latter generally indicates a finer division of layers.

Gangue: the waste or non-valuable minerals in the ore. In a copper ore, quartz is gangue. The word is pronounced as if spelled "gang".

Geodes: a hollow nodule or concretion, the cavity of which is later lined with crystals. The geode may ultimately become filled with the later minerals.

Gneiss: a banded crystalline rock with more or less well-developed cleavage but without the fissility (separation into thin sheets) of schist.

Gossan: a mass of iron oxide (limonite) that may cover a mineral deposit. The original rock has been weathered, many of the soluble minerals have been removed, and the weathered residue remains. Also called "iron hat".

Granite: a granular igneous rock composed essentially of quartz, feldspar, and mica. Commonly a part of the feldspar is plagioclase. The mica may be muscovite or biotite or both. Hornblende is a common, and augite an uncommon, component. Commercially almost all compact igneous rocks are called granite as distinguished from slate, sandstone, and marble. As used here, it applies to a particular rock as defined.

Granitoid: the suffix "oid" means "like". A granitoid rock therefore is one that is like a granite. It may be a granodiorite, diorite, or a similar rock. The term is useful as a generality.

Homogeneous: of the same kind or nature; consisting of similar parts, or of elements of a like nature; opposed to heterogeneous.

Hydrous: containing water chemically combined, as in hydrates and hydroxides.

Igneous: formed by crystallization from a molten state; said of the rocks of one of the great classes into which all rocks are divided, and contrasted with sedimentary and metamorphic rocks.

Iridescence: the exhibition of prismatic (rainbow) colors in the interior, or on the surface of a mineral. An oil film on water exhibits a play of colors and illustrates iridescence.

Iron Hat: see Gossan.

Magmatic Segregation: the process by which different types of igneous rocks are derived from a single parent magma (molten mass within the earth) or by which different parts of a single molten mass assume different compositions and textures as it solidifies. Also called Magmatic Differentiation.

Malleable: capable of being extended or shaped by beating with a hammer, as gold, silver, etc.

Matrix: the rock or earthy material containing a mineral or metallic ore. Some-
    times called Groundmass.

Megascopic: large enough to be distinguished with the naked eye; the opposite of
    Microscopic.

Metamorphism: any change in the texture or composition of a rock, often its indur-
    ation or solidification, especially by deformation and by a rise of temperature.

Micaceous: composed of thin scales, like a mass of tiny mica flakes.

Mineralogist: one who is versed in the science of minerals, or one who treats
    or discourses of the properties of mineral bodies.

Mohs Scale: an arbitrary scale, beginning with 1 and ending with 10, to designate
    varying degrees of hardness.   There is no relationship of the numbers; 4 is
    not necessarily twice as hard as 2.

Molecule: the smallest part of a substance that can exist separately and still
    retain its composition and characteristic properties; the smallest combination
    of atoms that will form a given chemical compound.

Nodular: having the shape of nodules (see Nodule).

Nodule: a small, roundish lump of mineral of irregular shape.

Nomenclature: the system of names used in a particular branch of knowledge or
    art, or by any school or individual (Webster).

Nucleus: a kernel; a central mass or point about which other matter is gathered,
    or to which an accretion is made.

Octahedral Form: in the isometric (cubic) system, a closed form of eight faces,
    each having equal intercepts on all three axes.

Oolite: a variety of limestone consisting of round grains like the roe of a fish;
    the name is derived from two Greek words which mean "egg-stone". Also ap-
    plies to any other material having a similar texture.

Oolitic: characteristic of, pertaining to, of the nature or texture of, or compos-
    ed of, oolite.   (See oolite).

Oxidized zone: that portion of an ore deposit which has been subjected to the ac-
    tion of surface waters carrying oxygen, carbon dioxide, etc. That zone in
    which sulfides have been removed or altered to oxides and carbonates.

Pervious: admitting passage; capable of being penetrated (Webster).

Pisolitic: consisting of rounded grains like peas or beans.

Polysynthetic Twinning: repeated twinning in which the crystal is made of thin
    lamellae alternately in reversed position. Plagioclase feldspar frequently
    has this twinning.

Primary Ores: those minerals and ores that retain their original form and composition, as original sulfides.

Pseudomorph: a crystal, or apparent crystal, of some mineral, having the outward form characteristic of another mineral, which has been replaced by substitution or by chemical alteration (see Replacement). Limonite is very frequently found in cubes, pseudomorphic after the original cubes of the pyrite from which it was derived.

Pyribole: a word that has been coined to describe a mineral that may be an amphibole or may be a pyroxene, and definite decision cannot be made with the resources at hand.

Quadrangle: the tract of country represented by one of the atlas sheets published by the United States Geological Survey. In densely populated areas each quadrangle measures 15' in latitude and 15' in longitude and is mapped on the scale of 1/62,500 (approximately 1 inch = 1 mile). Elsewhere the size is 30' x 30' and the scale of mapping is 1/125,000 (approximately 1 inch = 2 miles) except in a few sparsely settled regions when the size is 1° x 1° and the scale is 1/250,000 (approximately 1 inch = 4 miles). (Webster).

Quadrant: the quarter of a circle; an arc of 90°.

Replacements: the process by which one mineral or chemical substance takes the place of some earlier different substance, often preserving its structure or crystalline form. If the structure of the original mineral is preserved, the replacement is said to be a pseudomorph (see Pseudomorph).

Rhombohedral: a six-sided form whose faces are parallelograms. The resultant shape is that of a box which has been pushed out of shape by applying pressure at one corner.

Salic: a coined word, derived from silica-alumina, and pertains to rocks and minerals predominately composed of silica and alumina. Synonomous with "acidic", but salic is preferential. (See Acidic).

Schist: a crystalline rock that can be readily split or cleaved because of having a foliated or parallel structure, generally secondary and developed by shearing and recrystallization under pressure.

Secondary Mineral: a mineral resulting from the alteration of a primary mineral. Thus, original sulfides, by oxidation change to sulfates, carbonates, and oxides, and these by hydration become hydrous forms.

Sectile: capable of being severed by the knife with a smooth cut, but yet pulverizable. Distinguished from brittle and malleable.

Silica: an oxide of silicon, chemically known as silicon dioxide and having a chemical formula of SiO_2.

Silicate: of, or pertaining to, silica; containing silica, or partaking of its nature. Containing abundant quartz. Any mineral that contains silica is known as a silicate.

Stalactite: descending, columnar deposits, generally of calcite, formed on the roof of a cavern by the drip of mineral solutions. In the Oregon Caves, the "rock icicles" hanging from the roof are stalactites. (See Stalagmite).

Stalagmite: uprising, columnar deposits, generally of calcite, formed on the floor of a cavern by the drip of mineral solutions from the roof. Solutions dripping from the end of stalactites may form stalagmites beneath them. (See Stalactite).

Striations: very fine, parallel lines marking the surfaces or cleavage faces of minerals. Also, the channels or scratches made in rock scouring.

Tabular: having a shape like a table; flattened; laminated.

Tetrahedrons: a crystal form, in the isometric (cubic) system, enclosed by four faces having equal intercepts on all three axes. Each face is a triangle. A tetrahedron has four faces, - an octahedron has eight faces, while the faces of each are still triangular.

Twin Crystals: crystals, in which one or more parts, regularly arranged, are in reverse position with reference to the other part or parts. They often appear externally to consist of two or more crystals symmetrically united, and sometimes have the form of a cross or star. They also exhibit the composition in the reversed arrangement of part of the crystal faces, in the striae of the surface, and in reentrant angles; in certain cases the compound structure can only be surely detected by microscopic examination in polarized light. (See Carlsbad and Polysynthetic Twinning).

Vitreous: having the luster of broken glass, quartz, or calcite. The word is derived from vitreol, an old name for sulfuric acid. This acid has a vitreous luster.

Vug: a cavity in the rock, usually lined with a crystalline incrustation. It differs from an amygdule in that the amygdule was formed in a lava rock by expanding vapor (see Amygdule). A geode is usually somewhat spherical in shape, while a vug is more elongated. (See Geode).

Weathering: the group of processes whereby rocks on exposure to weathering change in character, decay, and finally crumble to soil. Weathering processes include: chemical action of air and rain water; plants and bacteria; mechanical action resulting from temperature changes.

BIBLIOGRAPHY
(with comments)

Selected references dealing with the general subject of mineralogy:

Allen, John Eliot, The Chromite Deposits of Oregon: Oregon State De-
    partment of Geology and Mineral Industries, bulletin no.8, 1938.

Brush, George J., (Penfield, S.L.), Manual of Determinative Mineral-
    ogy: John Wiley & Sons, Inc., New York, 1906, 16th ed.,($3 50).
    This book contains an introduction to blowpipe analysis; also 375
    figures and 75 tables for determination of mineral species by means
    of simple chemical experiment in the wet and dry way, and by physi-
    cal tests.

Butler, G. Montague, Pocket Handbook of Blowpipe Analysis: John Wiley &
    Sons, Inc., New York, ($1.25).
    The directions are elementary and complete; anyone with the proper
    instruments and this book sho ld be able to test most of the common
    minerals.

Butler, G. Montague, A Pocket Handbook of Minerals. John Wiley & Sons,
    Inc., New York, 2nd ed., ($3.00).
    This book gives all the details needed to identify most common min-
    erals.   Emphasis is placed on characteristic physical features.

Dake, Wilson, and Fleener, Minerals of the Quartz Family: McGraw-Hill
    Book Co , Inc , New York, 1938, ($2.50).
    Describes in detail a group of minerals of great interest.

Dana, Edward S., (Ford, W. E), Dana's Textbook of Mineralogy: John Wiley
    & Sons, Inc., 1932, 4th ed , ($5.50).
    A standard book of more value to one who has had some training in
    mineralogy

Eakle, Arthur, and Pabst, Ad lf Mineral Tables for the Determination of
    Minerals by Their Physical Prop rties. John Wiley & Sons, Inc , 2nd
    ed., ($1.50).
    Included are tables for mineral identification by megascopic means;
    the first elimination is based on streak.   A valuable book for the
    prospector.

English, George L., Getting Acquainted with Minerals: McGraw-Hill Book
    Co., Inc., 1934, New York, ($2.50).
    A splendid book for beginners written in a clear and interesting man-
    ner by an authority; it is highly recommended.

Fansett, George R. Field Tests for the Common Metals: Arizona Bureau of
    Mines, Mineral Technology Series no.36, Bulletin 136, University of
    Arizona, 1934.

Hawkins, A C. The Book of Minerals: John Wiley & Sons, Inc., New
    York, 1935 ($1.50).
    The important story of minerals and gems is well written and is rec-
    ommended for beginners.

Industrial Minerals and Rocks: Seely W. Mudd Series: American Institute
of Mining and Metallurgical Engineers, New York. 1937. ($5.00).
An edited volume, each chapter dealing with some industrial mineral
and written by the authority on that subject. The chapter discusses
occurrence, description, mining, milling and marketing, with excellent
bibliography. It is the only book of its kind and invaluable to anyone desiring information about industrial minerals.

Kemp, James Furman, Handbook of Rocks: Van Nostrand, New York, 1911, 5th
ed.. ($3.00).
A standard text on rocks to be used without the microscope. It has
not been revised recently, but is a valuable addition to any petrology
library.

Kraus, Edward Henry, and Holden, Edward Fuller, Gems and Gem Minerals:
McGraw-Hill Book Co., Inc., 1931, 2nd ed., ($3.00).
This book is not too technical for the amateur.

Libbey, F. W., Progress Report on Coos Bay Coal Fields: Oregon State
Department of Geology and Mineral Industries, Bulletin 2, 1938.

Loomis, Frederic Brewster, Field Book of Common Rocks and Minerals:
G. P. Putnam's Sons, New York, 1923,($3.50).
For field work, this book is an excellent guide; it has 74 plates
most of which are in full color; its 2" x 4" size handily fits the
pocket.

Mellor, J. W., Clay and Pottery Industries: C. Griffin & Co., London,15/8.

Mineralogist's Pocket Reference: The Colorado Assaying Co., Denver, Colo.
(Free).

Pirsson, Louis J., and Knopf, Adolf, Rocks and Rock Minerals: John Wiley
& Sons, Inc., New York, 1910, 2nd ed. ($3.50).
A manual for the student and the geologist, for identification of
minerals and rocks without the microscope. It is concisely written,
authoritative, and highly recommended.

Ries, Heinrich, Clays, Their Occurrence, Properties and Uses: John Wiley
& Sons, 1927, 3rd ed.

Rosevear, Francis Burt, Mineralogy Manual: The Porter Chemical Company,
Hagerstown, Maryland, 1936, ($1.00).
This is an excellent small manual for field use.

Rogers, Austin F., Introduction to the Study of Minerals: McGraw-Hill Book
Co., Inc., New York, 1937, 3rd ed. ($5.00).
A thorough, compact textbook for field or laboratory. Covers fields
of mineralogy, crystallography, blowpipe analysis, descriptive mineralogy, and the determination of minerals. Treats of the properties,
occurrences, association, and origin of minerals. Many determinations
require use of blowpipe or microscope, but information is also given
for megascopic field examination. It is highly recommended for the
student who wishes to study the science of minerals.

Smith, W. D., "Diatomaceous Earth in Oregon": Economic Geology, vol.27, no.8, 1932.

United States Bureau of Mines, Washington, D.C.   Various publications treating of minerals and the economics of production.   A catalogue of publications may be consulted in any library.   (These publications may be purchased from the Superintendent of Documents, Washington, D.C.)

United States Geological Survey, Washington, D.C.   (Same as for Bureau of Mines).

Wilson, Hewitt, Ceramics; Clay Technology.   McGraw-Hill Book Co., Inc., 1927.

Wilson, Hewitt, and Treasher, Ray C., Refractory Clays of Western Oregon: Oregon State Department of Geology and Mineral Industries, Bulletin no.6, 1938.

GOLD RUSH BOOKS

OREGON, USA

www.GoldRushBooks.com

More Books On Mining

Visit: www.goldrushbooks.com to order your copies or ask your favorite book seller to offer them.

Mining Books by Kerby Jackson

Gold Dust: Stories From Oregon's Mining Years

Oregon mining historian and prospector, Kerby Jackson, brings you a treasure trove of seventeen stories on Southern Oregon's rich history of gold prospecting, the prospectors and their discoveries, and the breathtaking areas they settled in and made homes. **5" X 8", 98 ppgs. Retail Price: $11.99**

The Golden Trail: More Stories From Oregon's Mining Years

In his follow-up to "Gold Dust: Stories of Oregon's Mining Years", this time around, Jackson brings us twelve tales from Oregon's Gold Rush, including the story about the first gold strike on Canyon Creek in Grant County, about the old timers who found gold by the pail full at the Victor Mine near Galice, how Iradel Bray discovered a rich ledge of gold on the Coquille River during the height of the Rogue River War, a tale of two elderly miners on the hunt for a lost mine in the Cascade Mountains, details about the discovery of the famous Armstrong Nugget and others. **5" X 8", 70 ppgs. Retail Price: $10.99**

Oregon Mining Books

Geology and Mineral Resources of Josephine County, Oregon

Unavailable since the 1970's, this important publication was originally compiled by the Oregon Department of Geology and Mineral Industries and includes important details on the economic geology and mineral resources of this important mining area in South Western Oregon. Included are notes on the history, geology and development of important mines, as well as insights into the mining of gold, copper, nickel, limestone, chromium and other minerals found in large quantities in Josephine County, Oregon. **8.5" X 11", 54 ppgs. Retail Price: $9.99**

Mines and Prospects of the Mount Reuben Mining District

Unavailable since 1947, this important publication was originally compiled by geologist Elton Youngberg of the Oregon Department of Geology and Mineral Industries and includes detailed descriptions, histories and the geology of the Mount Reuben Mining District in Josephine County, Oregon. Included are notes on the history, geology, development and assay statistics, as well as underground maps of all the major mines and prospects in the vicinity of this much neglected mining district. **8.5" X 11", 48 ppgs. Retail Price: $9.99**

The Granite Mining District

Notes on the history, geology and development of important mines in the well known Granite Mining District which is located in Grant County, Oregon. Some of the mines discussed include the Ajax, Blue Ribbon, Buffalo, Continental, Cougar-Independence, Magnolia, New York, Standard and the Tillicum. Also included are many rare maps pertaining to the mines in the area. **8.5" X 11", 48 ppgs. Retail Price: $9.99**

Ore Deposits of the Takilma and Waldo Mining Districts of Josephine County, Oregon

The Waldo and Takilma mining districts are most notable for the fact that the earliest large scale mining of placer gold and copper in Oregon took place in these two areas. Included are details about some of the earliest large gold mines in the state such as the Llano de Oro, High Gravel, Cameron, Platerica, Deep Gravel and others, as well as copper mines such as the famous Queen of Bronze mine, the Waldo, Lily and Cowboy mines. This volume also includes six maps and 20 original illustrations. **8.5" X 11", 74 ppgs. Retail Price: $9.99**

Metal Mines of Douglas, Coos and Curry Counties, Oregon

Oregon mining historian Kerby Jackson introduces us to a classic work on Oregon's mining history in this important re-issue of Bulletin 14C Volume 1, otherwise known as the Douglas, Coos & Curry Counties, Oregon Metal Mines Handbook. Unavailable since 1940, this important publication was originally compiled by the Oregon Department of Geology and Mineral Industries includes detailed descriptions, histories and the geology of over 250 metallic mineral mines and prospects in this rugged area of South West Oregon. **8.5" X 11", 158 ppgs. Retail Price: $19.99**

Metal Mines of Jackson County, Oregon

Unavailable since 1943, this important publication was originally compiled by the Oregon Department of Geology and Mineral Industries includes detailed descriptions, histories and the geology of over 450 metallic mineral mines and prospects in Jackson County, Oregon. Included are such famous gold mining areas as Gold Hill, Jacksonville, Sterling and the Upper Applegate. **8.5" X 11", 220 ppgs. Retail Price: $24.99**

Metal Mines of Josephine County, Oregon

Oregon mining historian Kerby Jackson introduces us to a classic work on Oregon's mining history in this important re-issue of Bulletin 14C, otherwise known as the Josephine County, Oregon Metal Mines Handbook. Unavailable since 1952, this important publication was originally compiled by the Oregon Department of Geology and Mineral Industries includes detailed descriptions, histories and the geology of over 500 metallic mineral mines and prospects in Josephine County, Oregon. **8.5" X 11", 250 ppgs. Retail Price: $24.99**

Metal Mines of North East Oregon

Oregon mining historian Kerby Jackson introduces us to a classic work on Oregon's mining history in this important re-issue of Bulletin 14A and 14B, otherwise known as the North East Oregon Metal Mines Handbook. Unavailable since 1941, this important publication was originally compiled by the Oregon Department of Geology and Mineral Industries and includes detailed descriptions, histories and the geology of over 750 metallic mineral mines and prospects in North Eastern Oregon. **8.5" X 11", 310 ppgs. Retail Price: $29.99**

Metal Mines of North West Oregon

Oregon mining historian Kerby Jackson introduces us to a classic work on Oregon's mining history in this important re-issue of Bulletin 14D, otherwise known as the North West Oregon Metal Mines Handbook. Unavailable since 1951, this important publication was originally compiled by the Oregon Department of Geology and Mineral Industries and includes detailed descriptions, histories and the geology of over 250 metallic mineral mines and prospects in North Western Oregon. **8.5" X 11", 182 ppgs. Retail Price: $19.99**

Mines and Prospects of Oregon

Mining historian Kerby Jackson introduces us to a classic mining work by the Oregon Bureau of Mines in this important re-issue of The Handbook of Mines and Prospects of Oregon. Unavailable since 1916, this publication includes important insights into hundreds of gold, silver, copper, coal, limestone and other mines that operated in the State of Oregon around the turn of the 19th Century. Included are not only geological details on early mines throughout Oregon, but also insights into their history, production, locations and in some cases, also included are rare maps of their underground workings. **8.5" X 11", 314 ppgs. Retail Price: $24.99**

Lode Gold of the Klamath Mountains of Northern California and South West Oregon

(See California Mining Books)

Mineral Resources of South West Oregon

Unavailable since 1914, this publication includes important insights into dozens of mines that once operated in South West Oregon, including the famous gold fields of Josephine and Jackson Counties, as well as the Coal Mines of Coos County. Included are not only geological details on early mines throughout South West Oregon, but also insights into their history, production and locations. **8.5" X 11", 154 ppgs. Retail Price: $11.99**

Chromite Mining in The Klamath Mountains of California and Oregon

(See California Mining Books)

Southern Oregon Mineral Wealth

Unavailable since 1904, this rare publication provides a unique snapshot into the mines that were operating in the area at the time. Included are not only geological details on early mines throughout South West Oregon, but also insights into their history, production and locations. Some of the mining areas include Grave Creek, Greenback, Wolf Creek, Jump Off Joe Creek, Granite Hill, Galice, Mount Reuben, Gold Hill, Galls Creek, Kane Creek, Sardine Creek, Birdseye Creek, Evans Creek, Foots Creek, Jacksonville, Ashland, the Applegate River, Waldo, Kerby and the Illinois River, Althouse and Sucker Creek, as well as insights into local copper mining and other topics. **8.5" X 11", 64 ppgs. Retail Price: $8.99**

Geology and Ore Deposits of the Takilma and Waldo Mining Districts

Unavailable since the 1933, this publication was originally compiled by the United States Geological Survey and includes details on gold and copper mining in the Takilma and Waldo Districts of Josephine County, Oregon. The Waldo and Takilma mining districts are most notable for the fact that the earliest large scale mining of placer gold and copper in Oregon took place in these two areas. Included in this report are details about some of the earliest large gold mines in the state such as the Llano de Oro, High Gravel, Cameron, Platerica, Deep Gravel and others, as well as copper mines such as the famous Queen of Bronze mine, the Waldo, Lily and Cowboy mines. In addition to geological examinations, insights are also provided into the production, day to day operations and early histories of these mines, as well as calculations of known mineral reserves in the area. This volume also includes six maps and 20 original illustrations. **8.5" X 11", 74 ppgs. Retail Price: $9.99**

Gold Mines of Oregon

Oregon mining historian Kerby Jackson introduces us to a classic work on Oregon's mining history in this important re-issue of Bulletin 61, otherwise known as "Gold and Silver In Oregon". Unavailable since 1968, this important publication was originally compiled by geologists Howard C. Brooks and Len Ramp of the Oregon Department of Geology and Mineral Industries and includes detailed descriptions, histories and the geology of over 450 gold mines Oregon. Included are notes on the history, geology and gold production statistics of all the major mining areas in Oregon including the Klamath Mountains, the Blue Mountains and the North Cascades. While gold is where you find it, as every miner knows, the path to success is to prospect for gold where it was previously found. **8.5" X 11", 344 ppgs. Retail Price: $24.99**

Mines and Mineral Resources of Curry County Oregon

Originally published in 1916, this important publication on Oregon Mining has not been available for nearly a century. Included are rare insights into the history, production and locations of dozens of gold mines in Curry County, Oregon, as well as detailed information on important Oregon mining districts in that area such as those at Agness, Bald Face Creek, Mule Creek, Boulder Creek, China Diggings, Collier Creek, Elk River, Gold Beach, Rock Creek, Sixes River and elsewhere. Particular attention is especially paid to the famous beach gold deposits of this portion of the Oregon Coast. **8.5" X 11", 140 ppgs. Retail Price: $11.99**

Chromite Mining in South West Oregon

Mining historian Kerby Jackson introduces us to a classic mining work in this important re-issue of the Oregon Department of Geology and Mineral Industries publication "Chromite in South West Oregon". Originally published in 1961, this important publication on Oregon Mining has not been available for nearly a century. Included are rare insights into the history, production and locations of nearly 300 chromite mines in South Western Oregon. **8.5" X 11", 184 ppgs. Retail Price: $14.99**

Mineral Resources of Douglas County Oregon

Mining historian Kerby Jackson introduces us to a classic mining work in this important re-issue of the Oregon Department of Geology and Mineral Industries publication "Geology and Mineral Resources of Douglas County, Oregon". Originally published in 1972, this important publication on Oregon Mining has not been available for nearly forty years. Included are rare insights into the geology, history, production and locations of numerous gold mines and other mining properties in Douglas County, Oregon. **8.5" X 11", 124 ppgs. Retail Price: $11.99**

Idaho Mining Books

Gold in Idaho

Unavailable since the 1940's, this publication was originally compiled by the Idaho Bureau of Mines and includes details on gold mining in Idaho. Included is not only raw data on gold production in Idaho, but also valuable insight into where gold may be found in Idaho, as well as practical information on the gold bearing rocks and other geological features that will assist those looking for placer and lode gold in the State of Idaho. This volume also includes thirteen gold maps that greatly enhance the practical usability of the information contained in this small book detailing where to find gold in Idaho. **8.5" X 11", 72 ppgs. Retail Price: $9.99**

Geology of the Couer D'Alene Mining District of Idaho

Unavailable since 1961, this publication was originally compiled by the Idaho Bureau of Mines and Geology and includes details on the mining of gold, silver and other minerals in the famous Coeur D'Alene Mining District in Northern Idaho. Included are details on the early history of the Coeur D'Alene Mining District, local tectonic settings, ore deposit features, information on the mineral belts of the Osburn Fault, as well as detailed information on the famous Bunker Hill Mine, the Dayrock Mine, Galena Mine, Lucky Friday Mine and the infamous Sunshine Mine. This volume also includes sixteen hard to find maps. **8.5" X 11", 70 ppgs. Retail Price: $9.99**

Utah Mining Books

Fluorite in Utah

Unavailable since 1954, this publication was originally compiled by the USGS, State of Utah and U.S. Atomic Energy Commission and details the mining of fluorspar, also known as fluorite in the State of Utah. Included are details on the geology and history of fluorspar (fluorite) mining in Utah, including details on where this unique gem mineral may be found in the State of Utah. **8.5" X 11", 60 ppgs. Retail Price: $8.99**

California Mining Books

The Tertiary Gravels of the Sierra Nevada of California

Mining historian Kerby Jackson introduces us to a classic mining work by Waldemar Lindgren in this important re-issue of The Tertiary Gravels of the Sierra Nevada of California. Unavailable since 1911, this publication includes details on the gold bearing ancient river channels of the famous Sierra Nevada region of California. **8.5" X 11", 282 ppgs. Retail Price: $19.99**

The Mother Lode Mining Region of California

Unavailable since 1900, this publication includes details on the gold mines of California's famous Mother Lode gold mining area. Included are details on the geology, history and important gold mines of the region, as well as insights into historic mining methods, mine timbering, mining machinery, mining bell signals and other details on how these mines operated. Also included are insights into the gold mines of the California Mother Lode that were in operation during the first sixty years of California's mining history. **8.5" X 11", 176 ppgs. Retail Price: $14.99**

Lode Gold of the Klamath Mountains of Northern California and South West Oregon

Unavailable since 1971, this publication was originally compiled by Preston E. Hotz and includes details on the lode mining districts of Oregon and California's Klamath Mountains. Included are details on the geology, history and important lode mines of the French Gulch, Deadwood, Whiskeytown, Shasta, Redding, Muletown, South Fork, Old Diggings, Dog Creek (Delta), Bully Choop (Indian Creek), Harrison Gulch, Hayfork, Minersville, Trinity Center, Canyon Creek, East Fork, New River, Denny, Liberty (Black Bear), Cecilville, Callahan, Yreka, Fort Jones and Happy Camp mining districts in California, as well as the Ashland, Rogue River, Applegate, Illinois River, Takilma, Greenback, Galice, Silver Peak, Myrtle Creek and Mule Creek districts of South Western Oregon. Also included are insights into the mineralization and other characteristics of this important mining region. **8.5" X 11", 100 ppgs. Retail Price: $10.99**

Mines and Mineral Resources of Shasta County, Siskiyou County, Trinity County, California

Unavailable since 1915, this publication was originally compiled by the California State Mining Bureau and includes details on the gold mines of this area of Northern California. Also included are insights into the mineralization and other characteristics of this important mining region, as well as the location of historic gold mines. **8.5" X 11", 204 ppgs. Retail Price: $19.99**

Geology of the Yreka Quadrangle, Siskiyou County, California

Unavailable since 1977, this publication was originally compiled by Preston E. Hotz and includes details on the geology of the Yreka Quadrangle of Siskiyou County, California. Also included are insights into the mineralization and other characteristics of this important mining region. **8.5" X 11", 78 ppgs. Retail Price: $7.99**

Mines of San Diego and Imperial Counties, California

Originally published in 1914, this important publication on California Mining has not been available for a century. This publication includes important information on the early gold mines of San Diego and Imperial County, which were some of the first gold fields mined in California by early Spanish and Mexican miners before the 49ers came on the scene. Included are not only details on early mining methods in the area, production statistics and geological information, but also the location of the early gold mines that helped make California "The Golden State". Also included are details on the mining of other minerals such as silver, lead, zinc, manganese, tungsten, vanadium, asbestos, barite, borax, cement, clay, dolomite, fluospar, gem stones, graphite, marble, salines, petroleum, stronium, talc and others. **8.5" X 11", 116 ppgs. Retail Price: $12.99**

Mines of Sierra County, California

Unavailable since 1920, this publication was originally compiled by the California State Mining Bureau and includes details on the gold mines of Sierra County, California. Also included are insights into the mineralization and other characteristics of this important mining region, as well as the location of historic gold mines. **8.5" X 11", 156 ppgs. Retail Price: $19.99**

Mines of Plumas County, California

Unavailable since 1918, this publication was originally compiled by the California State Mining Bureau and includes details on the gold mines of Plumas County, California. Also included are insights into the mineralization and other characteristics of this important mining region, as well as the location of historic gold mines. **8.5" X 11", 200 ppgs. Retail Price: $19.99**

Mines of El Dorado, Placer, Sacramento and Yuba Counties, California

Originally published in 1917, this important publication on California Mining has not been available for nearly a century. This publication includes important information on the early gold mines of El Dorado County, Placer County, Sacramento County and Yuba County, which were some of the first gold fields mined by the Forty-Niners during the California Gold Rush. Included are not only details on early mining methods in the area, production statistics and geological information, but also the location of the early gold mines that helped make California "The Golden State". Also included are insights into the early mining of chrome, copper and other minerals in this important mining area. **8.5" X 11", 204 ppgs. Retail Price: $19.99**

Mines of Los Angeles, Orange and Riverside Counties, California

Originally published in 1917, this important publication on California Mining has not been available for nearly a century. This publication includes important information on the early gold mines of Los Angeles County, Orange County and Riverside County, which were some of the first gold fields mined in California by early Spanish and Mexican miners before the 49ers came on the scene. Included are not only details on early mining methods in the area, production statistics and geological information, but also the location of the early gold mines that helped make California "The Golden State". **8.5" X 11", 146 ppgs. Retail Price: $12.99**

Mines of San Bernadino and Tulare Counties, California

Originally published in 1917, this important publication on California Mining has not been available for nearly a century. This publication includes important information on the early gold mines of San Bernadino and Tulare County, which were some of the first gold fields mined in California by early Spanish and Mexican miners before the 49ers came on the scene. Included are not only details on early mining methods in the area, production statistics and geological information, but also the location of the early gold mines that helped make California "The Golden State". Also included are details on the mining of other minerals such as copper, iron, lead, zinc, manganese, tungsten, vanadium, asbestos, barite, borax, cement, clay, dolomite, fluospar, gem stones, graphite, marble, salines, petroleum, stronium, talc and others. **8.5" X 11", 200 ppgs. Retail Price: $19.99**

Chromite Mining in The Klamath Mountains of California and Oregon

Unavailable since 1919, this publication was originally compiled by J.S. Diller of the United States Department of Geological Survey and includes details on the chromite mines of this area of Northern California and Southern Oregon. Also included are insights into the mineralization and other characteristics of this important mining region, as well as the location of historic mines. Also included are insights into chromite mining in Eastern Oregon and Montana. **8.5" X 11", 98 ppgs. Retail Price: $9.99**

Mines and Mining in Amador, Calaveras and Tuolumne Counties, California

Unavailable since 1915, this publication was originally compiled by William Tucker and includes details on the mines and mineral resources of this important California mining area. Included are details on the geology, history and important gold mines of the region, as well as insights into other local mineral resources such as asbestos, clay, copper, talc, limestone and others. Also included are insights into the mineralization and other characteristics of this important portion of California's Mother Lode mining region. **8.5" X 11", 198 ppgs. Retail Price: $14.99**

Alaska Mining Books

Ore Deposits of the Willow Creek Mining District, Alaska

Unavailable since 1954, this hard to find publication includes valuable insights into the Willow Creek Mining District near Hatcher Pass in Alaska. The publication includes insights into the history, geology and locations of the well known mines in the area, including the Gold Cord, Independence, Fern, Mabel, Lonesome, Snowbird, Schroff-O'Neil, High Grade, Marion Twin, Thorpe, Webfoot, Kelly-Willow, Lane, Holland and others. **8.5" X 11", 96 ppgs. Retail Price: $9.99**

More Mining Books

Prospecting and Developing A Small Mine

Topics covered include the classification of varying ores, how to take a proper ore sample, the proper reduction of ore samples, alluvial sampling, how to understand geology as it is applied to prospecting and mining, prospecting procedures, methods of ore treatment, the application of drilling and blasting in a small mine and other topics that the small scale miner will find of benefit. **8.5" X 11", 112 ppgs, Retail Price: $11.99**

Timbering For Small Underground Mines

Topics covered include the selection of caps and posts, the treatment of mine timbers, how to install mine timbers, repairing damaged timbers, use of drift supports, headboards, squeeze sets, ore chute construction, mine cribbing, square set timbering methods, the use of steel and concrete sets and other topics that the small underground miner will find of benefit. This volume also includes twenty eight illustrations depicting the proper construction of mine timbering and support systems that greatly enhance the practical usability of the information contained in this small book. **8.5" X 11", 88 ppgs. Retail Price: $10.99**

Timbering and Mining

A classic mining publication on Hard Rock Mining by W.H. Storms. Unavailable since 1909, this rare publication provides an in depth look at American methods of underground mine timbering and mining methods. Topics include the selection and preservation of mine timbers, drifting and drift sets, driving in running ground, structural steel in mine workings, timbering drifts in gravel mines, timbering methods for driving shafts, positioning drill holes in shafts, timbering stations at shafts, drainage, mining large ore bodies by means of open cuts or by the "Glory Hole" system, stoping out ore in flat or low lying veins, use of the "Caving System", stoping in swelling ground, how to stope out large ore bodies, Square Set timbering on the Comstock and its modifications by California miners, the construction of ore chutes, stoping ore bodies by use of the "Block System", how to work dangerous ground, information on the "Delprat System" of stoping without mine timbers, construction and use of headframes and much more. This volume provides a reference into not only practical methods of mining and timbering that may be employed in narrow vein mining by small miners today, but also rare insights into how mines were being worked at the turn of the 19th Century. **8.5" X 11", 288 ppgs. Retail Price: $24.99**

A Study of Ore Deposits For The Practical Miner

Mining historian Kerby Jackson introduces us to a classic mining publication on ore deposits by J.P. Wallace. First published in 1908, it has been unavailable for over a century. Included are important insights into the properties of minerals and their identification, on the occurrence and origin of gold, on gold alloys, insights into gold bearing sulfides such as pyrites and arsenopyrites, on gold bearing vanadium, gold and silver tellurides, lead and mercury tellurides, on silver ores, platinum and iridium, mercury ores, copper ores, lead ores, zinc ores, iron ores, chromium ores, manganese ores, nickel ores, tin ores, tungsten ores and others. Also included are facts regarding rock forming minerals, their composition and occurrences, on igneous, sedimentary, metamorphic and intrusive rocks, as well as how they are geologically disturbed by dikes, flows and faults, as well as the effects of these geologic actions and why they are important to the miner. Written specifically with the common miner and prospector in mind, the book will help to unlock the earth's hidden wealth for you and is written in a simple and concise language that anyone can understand. **8.5" X 11", 366 ppgs. Retail Price: $24.99**